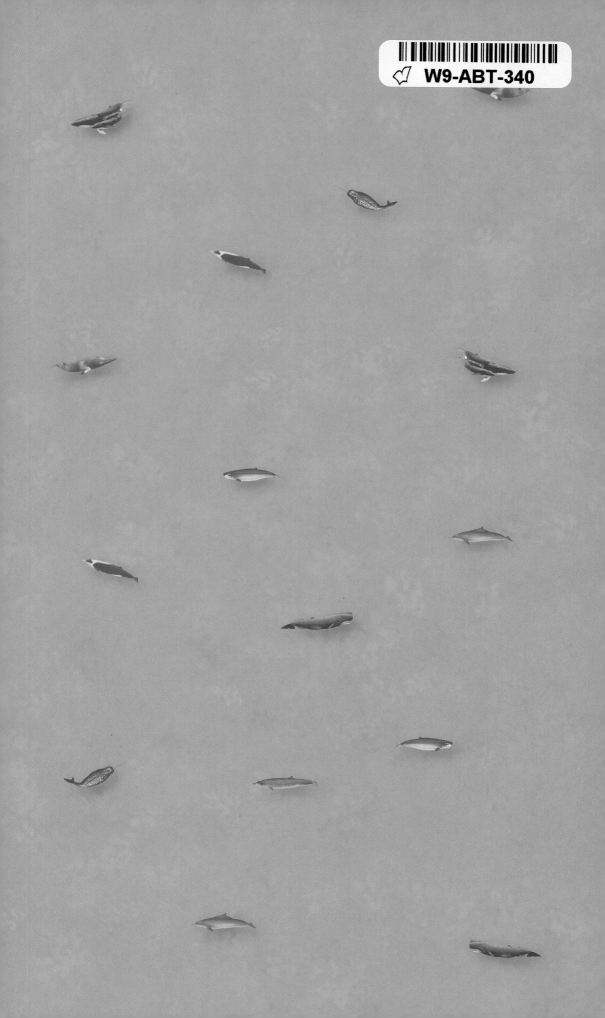

The Nature Company Guides

WHALES,
DOLPHINS & PORPOISES

The Nature Company Guides

WHALES,
DOLPHINS & PORPOISES

MARK CARWARDINE, ERICH HOYT,
R. EWAN FORDYCE, PETER GILL

CONSULTANT EDITORS
MARK CARWARDINE AND ERICH HOYT

THE
NATURE
COMPANY

TIME
LIFE
BOOKS

The Nature Company Guides are published by Time-Life Books

Conceived and produced by Weldon Owen Pty Limited
43 Victoria Street, McMahons Point, NSW, 2060, Australia
A member of the Weldon Owen Group of Companies
Sydney • San Francisco
Copyright 1998 © US Weldon Owen Inc.
Copyright 1998 © Weldon Owen Pty Limited

The Nature Company owes its vision to the world's great naturalists:
Charles Darwin, Henry David Thoreau, John Muir, David Brower,
Rachel Carson, Jacques Cousteau, and many others.
Through their inspiration, we are dedicated to providing products and
experiences which encourage the joyous observation, understanding, and
appreciation of nature. We do not advocate, and will not allow to be sold in
our stores, any products that result from the killing of wild animals for trophy
purposes. Seashells, butterflies, furs, and mounted animal specimens fall into
this category. Our goal is to provide you with products, insights, and
experiences which kindle your own sense of wonder and help you to feel
good about the world in which you live.
For a copy of The Nature Company mail-order catalog, or to learn the
location of the store nearest you, please call 1-800-227-1114.

Dedicated to the memory of Stephen Leatherwood, who inspired and
helped so many people in the study and conservation of cetaceans.

THE NATURE COMPANY
Priscilla Wrubel, Ed Strobin, Steve Manning,
Georganne Papac, Tracy Fortini

TIME LIFE CUSTOM PUBLISHING
Time-Life Books is a division of Time Life Inc.
Time-Life is a trademark of Time Warner Inc. U.S.A.

VICE PRESIDENT AND PUBLISHER: Terry Newell
EDITORIAL DIRECTOR: Donia A. Steele
DIRECTOR OF NEW PRODUCT DEVELOPMENT: Quentin McAndrew
DIRECTOR OF SALES: Neil Levin
DIRECTOR OF FINANCIAL OPERATIONS: J. Brian Birky

THE NATURE COMPANY GUIDES
PUBLISHER: Sheena Coupe
ASSOCIATE PUBLISHER: Lynn Humphries
PROJECT EDITORS: Helen Bateman, Elizabeth Connolly, Kathy Gerrard
COPY EDITOR: Lynn Cole
EDITORIAL ASSISTANTS: Vesna Radojcic, Shona Ritchie
DESIGNERS: Clive Collins, Lena Lowe
JACKET DESIGN: John Bull
PICTURE RESEARCH: Annette Crueger
ILLUSTRATIONS: Martin Camm, Marjorie Crosby-Fairall,
Ray Grinaway, Gino Hasler,
Roger Swainston, Genevieve Wallace
PRODUCTION MANAGER: Caroline Webber
PRODUCTION ASSISTANT: Kylie Lawson

Library of Congress Cataloging–in–Publication Data
    Whales, dolphins & porpoises/Mark Carwardine … [et al.],
consultant editors, Mark Carwardine and Erich Hoyt.
    p. cm. — (The Nature Company guides)
    Includes bibliographical references and index.
    ISBN 0–7835–5284–X
    1. Cetacea. I. Carwardine, Mark.   II Hoyt, Erich.
III. Series: Nature Company guide.
QL737.C4W4417 1998   97–37812
599.5—dc21   CIP

Color Reproduction by Leo Reprographic Ltd, Hong Kong
Printed by Leefung-Asco Printers
Printed in China

A Weldon Owen Production

The sea never changes and its works,

for all the talk of men, are wrapped in mystery.

Typhoon

JOSEPH CONRAD (1857–1924), Polish-born English essayist

Contents

FOREWORD

Those of us who love to travel and observe animals in their natural habitat have all, at one time or another, sat around a campfire and told stories of our adventures, and nominated which creature we would love to come back as in another life. Some of us pick eagles, hawks, elephants, lions, or tigers. My answer for the last 20 years has been, "I'd love to be a whale."

How fantastic it would be to be able to live underwater and experience that extraordinary environment, equipped with a body that is comfortable lolling on the surface of a great ocean, breathing the air, and seeing the stars, and yet equally at home diving thousands of feet into the deep, dark ocean. Imagine seeing the bizarre creatures that inhabit these depths. More remarkable still would be the ability to swim the many hundreds of miles from the polar oceans with their icebergs, abundant food, and days of constant sunlight, to the warm waters of the tropics. Months would pass there, spent frolicking, courting, giving birth, swimming with tropical fish—and all without eating. And then, incredibly, you would swim back to the polar regions before dining again.

There is still so much to learn about—and from—whales, dolphins, and porpoises. We can all play a part in that learning process, helping support whale research by taking advantage of the ecotourism options that are now available in the areas where whales are most easily seen. These spots exist all around the world, as this book shows. Once you have read this Guide and hopefully been inspired to go on a whale-watching trip, you won't fail to appreciate the majesty and importance of these great creatures.

PRISCILLA WRUBEL
Founder, The Nature Company

INTRODUCTION

Within, say, 20 feet of a whale your perspective changes. It's bigger than you think. In fact, it takes up all peripheral vision. You see individual barnacles or bits of skin sloughing off its back and little fish eating it, scratches, teeth marks, shark bites, harpoon wounds, and spindly little hairs on its chin. If underwater, you see its eyes bugging out, as if on stalks. At times it shivers in anticipation of contact—its skin is so sensitive. You see the intimate interactions of mother and calf, or the awesome ferocity and heat of breeding. It's the getting close—or letting the living whale get close to us—that has so changed our view of these animals.

I've been fortunate to be one of a cadre of researchers that studies living whales at sea. Not too long ago, this was considered all but impossible—until in different parts of the world, a handful of researchers began to use small boats to approach and observe whales. This engagement of the living, wild animal by curious people opened the door to understanding the nature of the whale.

More than 80 species of whale, dolphin, and porpoise inhabit our oceans and rivers, each with a different natural history. Some live in complex societies with sophisticated communications—their calls may be heard over thousands of miles of deep ocean basins. Some whales sing songs. Some migrate across oceans. Some dive to remarkable depths, functioning almost entirely in the pitch dark. Whales know the ocean in ways that are literally beyond our imagination. This *Nature Company Guide* shares insight into the lives of whales and dolphins, and explores the three-quarters of this planet that is their home.

James Darling

JAMES DARLING
Director, West Coast Whale Research Foundation

We need another and a wiser and perhaps
a more mystical concept of animals.

The Outermost House
HENRY BESTON (1888–1968), American writer

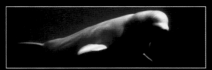

CHAPTER ONE

UNDERSTANDING WHALES, DOLPHINS, *and* PORPOISES

THE WORLD *of* WHALES, DOLPHINS, *and* PORPOISES

Whales, dolphins, and porpoises belong to a single group of marine mammals known as the cetaceans.

A total of 81 different species are currently recognized by whale experts although, as research progresses, new ones are still being discovered. A strange skull, for example, was found in the Juan Fernández Islands, off the coast of Chile, in June 1986. After nearly a decade of painstaking examination and scientific discussion, a team of whale experts finally concluded that this discovery was new to science. It was named Bahamonde's beaked whale in 1996, making it the most recent cetacean species to be formally recognized. Although believed to be a living species, no live representatives of this whale have yet been seen.

Meanwhile, a number of other cetacean species have been split in recent years. In 1995, for example, the common dolphin was officially separated into two distinct species, now known as the long-beaked common dolphin and the short-beaked common dolphin.

INFINITE VARIETY
Inevitably, whales, dolphins, and porpoises share many features in common. Yet they also come in an impressive variety of shapes, sizes, and colors; live in many different marine and freshwater habitats; and have developed a bewildering variety of adaptations for survival in their underwater world.

Some live in shallow water close to shore, or in major rivers and estuaries, while others live so far out to sea that they probably never set eyes on land from the day they are born until the day they die. Some are fairly common and widespread, while others are on the verge of extinction.

They range in size from several small dolphins and porpoises, as little as 4 feet (1.2 m) in length, to the enormous blue whale, which can grow to more than 98 feet (30 m) in length,

almost as long as a Boeing 737. Some species are brightly colored, with a motley collection of spots and stripes, several have striking black-and-white markings, while others are a relatively drab brown or gray. Some are long and slender, others short and robust. Some have tall, scythe-shaped dorsal fins; other species have much smaller, triangular fins; and several have no fins at all. There are even variations among individuals of the same species: between males and females, youngsters and adults, and among populations in different parts of the world.

SPECIAL APPEAL
There is something special and particularly appealing about cetaceans. This is difficult to put into words, and impossible to prove, yet it is a feeling shared by a great many people.

PHYSICAL VARIATION *The short-beaked common dolphin (top). Atlantic spotted dolphins (right). At the far right, is the curved fin (top) of a short-finned pilot whale, and the small dorsal fin (bottom) of a humpback whale.*

CHAMPIONS *Sperm whales, such as the one shown above, have made the longest and deepest dives ever recorded.*

Some claim it is because of their apparent intelligence; others are in awe that air-breathing mammals, like us, are able to thrive in such an alien underwater world. Some people are inspired by their ability to explore places out of our reach, experience things we will never experience, and see things we will never see. Others see them as we would like to see ourselves—free, graceful, compassionate, peaceful, and full of energy.

Undoubtedly, the amazing sense of mystery surrounding these enigmatic creatures is itself a major part of their appeal. We now know that many of our early assumptions about them were wrong or, at least, were not entirely accurate. Yet the more we learn about these incredible creatures, the more intriguing their story becomes.

RECORD BREAKERS

Perfectly adapted to life underwater, whales, dolphins, and porpoises are the record breakers of the animal kingdom. Here are a few examples:

The blue whale is the largest animal on Earth. The heaviest ever recorded was a female weighing 209 tons (190 tonnes), caught in the Southern Ocean in 1947. The longest was another female, also landed in the Southern Ocean, in 1909, measuring 110 feet 2 inches (33.5 m) from the tip of her snout to the end of her tail.

The low-frequency pulses made by blue whales and fin whales, when communicating with members of their own species across enormous stretches of ocean, have been measured at up to 188 decibels—the loudest sounds emitted by any living source.

The longest and most complex songs in the animal kingdom are sung by male humpback whales. Each song can last for half an hour or more and consists of several main components.

The humpback whale also undertakes the longest documented migration of any individual mammal (a record previously believed to be held by the gray whale). One humpback, for example, was observed at its feeding grounds near the Antarctic Peninsula and, less than five months later, was seen again at its breeding grounds off the coast of Colombia. The shortest swimming distance between these two locations is 5,176 miles (8,334 km).

The sperm whale is believed to dive deeper than any other mammal. The deepest known dive was 6,560 feet (2,000 m), recorded in 1991. It was made by a male sperm whale diving off the coast of Dominica, in the Caribbean. Indirect evidence suggests that sperm whales may be able to dive to depths of at least 10,000 feet (3,000 m). The record for the longest dive by any mammal is also held by the sperm whale; on 11 November 1983, biologists working in the southwest Caribbean listened to five sperm whales clicking underwater during a dive lasting an astonishing 2 hours 18 minutes.

THE OCEAN REALM

With seas and oceans covering more than two-thirds of its surface, planet Earth might more appropriately be called planet Water.

Seas and oceans are often treated as a single entity, like tropical rain forests, deserts, or mountains. But they contain many distinct habitats as different from one another as grassland is from woodland. These include rocky or sandy coastlines, estuaries, mud flats, mangrove swamps, coral reefs, salt marshes, kelp forests, open ocean, and ocean depths.

In the scientific world, the marine environment is usually subdivided into four major ecological zones. First is the intertidal or littoral zone, which is underwater at high tide and exposed to the air at low tide. Sandy beaches, salt marshes, mud flats, and mangrove swamps are all intertidal habitats; they are rarely used by cetaceans themselves, but are critically important as the breeding grounds of fish and other prey species.

Next is the continental shelf or sublittoral zone, the shallow area of sea closest to a continent. The continental shelf typically extends seaward to a distance of about 45 miles (70 km), but tends to be wider off low-lying regions and narrower off mountainous regions. It dips gently from the shoreline to an average depth of about 650 feet (200 m) and is a rich source of food for cetaceans and other predators, especially in temperate zones.

The continental slope, or bathyal zone, comes next.

This dips much more steeply from the edge of the continental shelf to the ocean bottom. It typically ends at a depth of about 3,500 feet (1,100 m) or more.

The ocean bottom itself, or abyssal plain as it is often known, is the fourth zone. It is extremely flat—more so than almost anywhere else on Earth—and occupies more than 40 percent of the oceanic area. Its depth is highly variable from region to region, but averages about 13,000 feet (4,000 m). Few animals live in its inhospitable, cold, dark waters. In some parts of the world, the abyssal plain is punctuated by ocean trenches and troughs that reach phenomenal depths. The deepest point is the Marianas Trench in the western Pacific, which drops to at least 35,800 feet (11,000 m).

DEPTH CHANGES

Even within these major ecological zones, there are different habitats at different depths. For example, the

CHANGING HABITATS *Giant kelp forests (above right) off Catalina Island, California, with a brilliant golden orange fish, a garibaldi, swimming among the tall kelp growth. A humpback in the cold waters of the Antarctic (above left).*

deeper you move down the water column, the darker it gets: the different colors of sunlight disappear, one by one, as depth increases (red is absorbed first, followed by orange, then yellow, green, and, ultimately, blue). As well as affecting visibility, this also influences the distribution of animal life in the water column. Where there is light, there are minute plants to sustain minute animals, which, in turn, are preyed on by larger animals.

Meanwhile, deeper water also means greater pressure: for every 33 feet (10 m) you descend, pressure increases by about 1 atmosphere. This means that a sperm whale diving to a depth of, say, 6,500 feet (2,000 m) has to be able to withstand pressures of as much as 200 atmospheres.

THE WARM, SHALLOW WATERS *of the Atlantic, in the Bahamas, are perfect for Atlantic spotted dolphins (above).*

Water temperature is also affected by depth and has a major impact on many biological, chemical, and physical processes in the sea. Oceans tend to have different layers of temperature, though the precise structure varies seasonally and from region to region; the temperature change from one layer to another frequently acts as a substantial barrier to different marine creatures. There is also great variation in the surface temperature of the sea in various parts of the world. It is highest at the equator, where it can sometimes reach 105 degrees Fahrenheit (40° C) in shallow tropical lagoons, and lowest in the high polar regions, where it can fall as low as 28 degrees Fahrenheit (−1.9° C).

CETACEAN HABITATS

Unfortunately, we know very little about the specific habitat requirements of most whales, dolphins, and porpoises. It is clear that each species, or population, tends to choose waters of a particular depth and temperature range but, within these requirements, many other factors have to be taken into consideration.

Their distribution is often affected by the pattern of major ocean currents. The Earth's rotation causes surface currents to flow clockwise in the Northern Hemisphere, which results in warm tropical waters moving farther north along the east coasts of continental land masses. So the distribution of warm-water species frequently extends farther north than might otherwise be expected. In the Southern Hemisphere, surface currents move anticlockwise, so cold polar waters move farther north along the west coasts of continental land masses. As a result, cold-water species frequently range surprisingly close to the equator.

Distribution is also affected by a process known as upwelling. Areas of upwelling are caused by a combination of winds and currents, which bring cold, nutrient-rich water from the ocean depths up to the surface of the water. Food is particularly abundant in these areas, which, inevitably, tend to attract large numbers of cetaceans and other top predators.

AWAY FROM THE OCEAN

The marine environment is not the only one inhabited by cetaceans. Four river dolphins, for example, live in some of the largest, muddiest rivers in Asia and South America. The Amazon River dolphin, in particular, can be found more than 2,000 miles (3,200 km) inland in some areas.

Several other cetaceans regularly inhabit fresh water. Some travel freely between river and sea but often there are separate riverine populations that seem to spend all their lives in fresh water; for example, finless porpoises (below) in the Yangtze River, China; tucuxis in the Amazon and Orinoco rivers, South America; and Irrawaddy dolphins in several major river systems of the Indo-Pacific.

IS IT *a* WHALE? *a* DOLPHIN? *or a* PORPOISE?

The common names used for animals can be confusing at the best of times,

but whales, dolphins, and porpoises are in a league of their own.

The words "whale," "dolphin," and "porpoise" are misleading. They have no real scientific basis and are the cause of much confusion. In theory, whales are the largest of the cetaceans, dolphins are of medium size, and porpoises are the smallest. But this does not always work: some whales are smaller than the largest dolphins, and some dolphins are smaller than the largest porpoises.

The situation is even more complex in North America, where small cetaceans of all kinds are commonly referred to as porpoises.

There are six "whales," in particular, that should really be called "dolphins." Despite their names, the killer whale, short-finned pilot whale, long-finned pilot whale, false killer whale, melon-headed whale, and pygmy killer whale are all members of the dolphin family, Delphinidae. It could be argued that the killer whale would more appropriately be called the killer dolphin (although most experts prefer to call it the orca, since it does not really deserve to be called "killer" at all). To add to the confusion, these six species are often grouped together as the "blackfish," which is particularly strange since not all are black and, of course, none is a fish. Ironically, there is even a "dolphinfish" that is really a white-fleshed saltwater fish.

Other species have equally confusing names, but for different reasons. The two right whale dolphins were named after northern and southern right whales (since all four species have finless backs). Curiously, not all white-beaked dolphins have white beaks. The Irrawaddy dolphin lives in the Brahmaputra, Ganges, Mekong, and Makakam rivers, and in the coastal waters of southeast Asia, as well as in the Irrawaddy River. And Indo-Pacific hump-backed dolphins living east and south of Sumatra, in Indonesia, do not have humps.

At the same time, many cetaceans are known by umpteen common names in at least as many languages. Thus the long-snouted spinner dolphin, spinner dolphin, long-beaked dolphin, rollover, and longsnout are all the same species. Likewise, the fin whale is also known as the herring whale, the finback, finner, common rorqual, and razorback. Even on the few occasions when everyone agrees on a name, there can be disagreement over the spelling, as in bottlenose, bottlenosed, and bottle-nose dolphin.

CLASSIFYING CETACEANS

The solution is to think of modern cetaceans in terms of two distinct groups, instead of three: the toothed whales, or odontocetes, which possess teeth; and the baleen whales,

VARIATION *The orca (left), or killer whale, seen here displaying its teeth, is not a whale, but a dolphin in the odontocetes group. The huge southern right whale (above) filters small food items through its prominent comb-like baleen plates.*

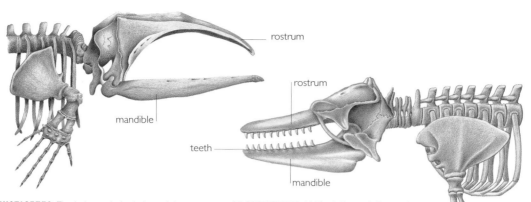

MYSTICETES *The baleen whale skeleton (above left) shows the arch-shaped rostrum that forms an anchor for the baleen plates in these whales.*

ODONTOCETES *Unlike baleen whales, teeth are attached to the rostrum and mandible in toothed whales, as shown in this skeleton (above right).*

or mysticetes, which do not. These groups have a strong scientific basis, and avoid all the confusion normally associated with "whales," "dolphins," and "porpoises."

The vast majority of cetaceans are odontocetes. There are 70 species in all, including all the oceanic dolphins, river dolphins, porpoises, beaked whales, and sperm whales, as well as the narwhal and beluga. The number, size, and shape of their teeth varies enormously. The long-snouted spinner dolphin, for example, has the most teeth, with the exact number varying from 172 to 252. At the other extreme, a number of species have only two teeth and, in many females, even these do not erupt (so they appear to have no teeth at all).

In the main, odontocetes feed on fish and squid, although some also take a variety of crustaceans, and a few take marine mammals.

The mysticetes comprise the remaining 11 species and make up for their lack of numbers by including many of the large and most popular whales, such as the blue, gray, humpback, and bowhead. Instead of teeth, they have hundreds of furry, comb-like baleen plates, often referred to as whalebone, which hang from their upper jaw. These are tightly packed inside the whales' mouth, and have stiff hairs that form a sieve-like structure to filter food out of the sea water. Mysticetes feed mainly on small schooling fish or crustaceans, such as krill and copepods.

There are other, more subtle, differences between odontocetes and mysticetes. Toothed whales, for example, are recognizable by their single blowholes, while baleen whales have two blowholes side by side.

WHALES, DOLPHINS, AND PORPOISES

FAMILY		No. of SPECIES
Baleen whales (Mysticetes)		
Balaenidae	right whales and bowhead whale	3
Neobalaenidae	pygmy right whale	1
Eschrichtiidae	gray whale	1
Balaenopteridae	rorqual whales	6
		Total 11
Toothed whales (Odontocetes)		
Kogiidae	pygmy and dwarf sperm whales	2
Physeteridae	sperm whale	1
Monodontidae	narwhal and beluga	2
Ziphiidae	beaked whales	21
Delphinidae	oceanic dolphins	33
Iniidae	boto (Amazon River dolphin)	1
Pontoporiidae	Yangtze River dolphin and franciscana	2
Platanistidae	Indus and Ganges River dolphins	2
Phocoenidae	porpoises	6
		Total 70

BOTO, OR AMAZON RIVER DOLPHINS, *(above), found only in the Amazon and Orinoco River basins, are toothed whales.*

NAMING WHALES, DOLPHINS, *and* PORPOISES

Naming whales, and working out which ones are closely related, is a challenging but very important branch of whale research.

Mention *Balaenoptera musculus* to most people and the chances are they will have no idea what you are talking about. But the eyes of a group of whale biologists will light up immediately. The reason is simple—*Balaenoptera musculus* is the scientific name for a blue whale.

Admittedly, "blue whale" is the more appealing of the two names. It is certainly the easier one to remember, so it is not surprising that most people use it in preference to *Balaenoptera musculus*. What causes endless problems is that species often have common names in many languages. The blue whale, for example, has several alternative names in English, as well as lots of foreign ones: *shiro nagasu kujira* in Japanese, *blåhval* in Norwegian, *ballena azul* in Spanish, *baleine bleue* in French, and so on. Its scientific name, on the other hand, is used by whale biologists throughout the world, whatever language they speak. Since no species has more than one scientific name, and no two species have exactly the same scientific name, there is no room for confusion. So, biologists in different parts of the world can always be sure they are talking about exactly the same kind of whale.

The scientific name is Latinized and is written in italics or underlined. It normally consists of two different words: the first begins with a capital letter and identifies the genus, while the second begins with a lower-case letter and identifies the species. There might also be a third word to identify the sub-species (if there is one).

THE PRINCIPLES OF TAXONOMY

Another reason for assigning a unique scientific name to each species is to place it within a structure of relationships in the animal kingdom. Biologists classify all living things by arranging them in groups, according to their similarities and differences. This very specialized science is known as taxonomy.

It is rather like working out an enormously complex family tree. The basic unit is the species, which is defined as a population of animals whose members do not freely interbreed with members of other populations. A genus is simply a group of closely related species. In the same way, a group of closely related genera (plural of "genus") forms a family; then closely related families are grouped into orders, closely related orders into classes, and so on.

This is the theory of classification, and it works very well—most of the time.

BRINGING ORDER *Swedish naturalist Carl Linnaeus (left) evolved the system of classification still used today. A southern right whale (above left).*

CLASSIFYING THE MINKE WHALE AND HUMPBACK WHALE

Minke and humpback whales are classified in the following way:

	Minke	Humpback
Kingdom	Animalia (animals)	Animalia (animals)
Phylum	Chordata (chordates)	Chordata (chordates)
Subphylum	Vertebrata (vertebrates)	Vertebrata (vertebrates)
Class	Mammalia (mammals)	Mammalia (mammals)
Order	Cetacea (cetaceans)	Cetacea (cetaceans)
Suborder	Mysticeti (baleen whales)	Mysticeti (baleen whales)
Family	Balaenopteridae (rorquals)	Balaenopteridae (rorquals)
Genus	*Balaeonoptera*	*Megaptera*
Species	*acutorostrata*	*novaeangliae*

In practice, though, scientific names and groupings sometimes have to be changed as new information comes to light, or when biologists disagree over the details of a particular grouping. It is also necessary, in some cases, to add extra groupings (subspecies, subfamilies, and suborders) to deal with more complex relationships within the system.

It sometimes resulted in unrelated species being linked, and closely related species being kept apart.

These days, classification of cetaceans is highly sophisticated. Taxonomists gather information from many disciplines, including physiology, behavioral biology, ecology, paleontology, and even biochemistry.

THE SCIENTIFIC NAME Balaenoptera musculus *immediately identifies the blue whale (above) as a filter feeder.*

CLASSIFYING CETACEANS

Taxonomy's real challenge is in establishing which animals are most closely related. All of the whales, dolphins, and porpoises are related to some extent, but clever detective work is needed to determine the different levels of these particular relationships.

Early taxonomists classified cetaceans almost entirely on their external appearance, but this was never a satisfactory system. At its crudest, it was tantamount to assuming that all men with gray beards and bald heads must be related.

DNA CLASSIFICATION

Recent work in the field of biochemical research, especially on DNA (deoxyribonucleic acid) analysis, has had a particularly dramatic impact on the classification of the animal kingdom (see also p. 31). In essence, it enables scientists to measure the level of relationships among species in an extraordinarily precise way.

DNA is the basic genetic material found in all animals. It is a kind of instruction manual for the design and assembly of the body's

proteins. Every cell in an animal's body contains an exact replica of this manual and almost all the "pages" go to making it what it is—a human, a dog, a humpback whale, a long-snouted spinner dolphin, and so on. The small number of "pages" that are left help to distinguish one individual from another—just as one person's fingerprints are different from everyone else's, no two animals have exactly the same DNA.

Making sense of all the similarities and differences in DNA is a highly complex process, but it has already resolved a number of long-standing uncertainties in the classification of whales, dolphins, and porpoises and other wildlife.

FEATURES *The humpback whale (left) breaches more frequently than other baleen whales. The minke whale (below) takes its species name,* acutorostrata, *from its slim, pointed head.*

Order marches with weighty and measured strides; disorder is always in a hurry.

Maxims,
NAPOLEON BONAPARTE
(1769–1821), Emperor of France

IMAGINARY CETACEANS

Whales and dolphins have captured the imaginations of artists and storytellers for thousands of years.

From classical Greek mythology to modern-day films such as *The Big Blue* and *Free Willy*, cetaceans have long been the stars of mythology, folklore, religion, and, most recently, popular entertainment. Early Icelanders used to tell stories about horse whales, boar whales, and fearsome red-headed whales, which reportedly roamed the seas in enormous herds and destroyed every ship and person in their path. The male narwhal was touted as the mythical unicorn, a white horse with a horn growing out of its forehead. Until European whalers uncovered the truth, enterprising northern traders marketed the narwhal's tusk as the mythical unicorn's horn. And even today, on the other side of the world, people living in the Amazon tell the story of river dolphins disguised as men who, during village fiestas, come ashore to woo the local girls.

SEA UNICORN *Narwhal tusks used to be sold as the horns of the mythical unicorn.*

JONAH AND THE WHALE

The best known whale in the Bible swallowed and then spat out the prophet Jonah. In real life, of course, Jonah could never have survived in the belly of a whale, but the myth has persisted in the creations of novelists, poets, painters, and sculptors ever since.

The story describes how Jonah was instructed by God to preach to the Gentiles in Nineveh, the Assyrian capital. Jonah balked at the mission—fearing the reaction of the luxury-loving Assyrians—and fled to Joppa where he boarded a ship. But the ship was caught in a terrible storm and, believing it to be a sign of God's anger, the crew threw Jonah overboard. He asked God to have mercy on him and immediately "the sea ceased from her raging" and he was swallowed by a "great fish" (which has long been presumed to be a sperm whale). Jonah remained inside the whale's belly for three days and three nights—until the whale "vomited out Jonah upon the dry land." Having learned his lesson, he promptly did as God had commanded.

FLIPPER THE DOLPHIN

The first cetacean to become a TV star was a bottlenose dolphin, made famous by the

BELLY OF THE WHALE *Illumination from an early fifteenth-century Bible (left) depicting Jonah and the whale.*

1960s TV series *Flipper*. The series portrayed our modern image of the "friendly" dolphin—whenever people were in trouble and needed help, Flipper and his friend, a young boy, were always there to save the day.

Six dolphins were actually used for the part of Flipper, and one of them did not enjoy filming. Ironically, considering the role she was playing, she was bad-tempered and belligerent for much of the time. Named Patty, she would rush at people in the water—actors included—trying to give them a fright and generally demonstrating her displeasure. One day, her trainer decided to teach her a lesson and, as she swam past, he gave her a thump on the back. With no immediate reaction, she continued swimming to the other side of the pool, turned slowly, and then accelerated toward him like a rocket; the next thing the

You can't depend on your judgement when your imagination is out of focus.

Notebook,
MARK TWAIN (1835–1910),
American writer and humorist

trainer remembers, he was coming round with a concussion in the local hospital.

MOBY DICK
Opening with the now-famous words "Call me Ishmael," *Moby Dick* chronicles the heyday of American whaling, when hundreds of thousands of sperm whales were slaughtered for their pure, odorless oil. It tells the story of Captain Ahab, who loses a leg in a battle with Moby Dick, the biggest and most majestic sperm whale of them all.

His leg replaced by a gleaming ivory jaw, Ahab becomes obsessed with killing the whale and hunts it all over the world until, eventually, Moby sinks the ship and Ahab goes down with it. The narrator is the young Ishmael, sole survivor of the deadly confrontation, and he describes whale species, whaling ships, harpooning techniques, and other aspects of the industry in great detail.

It wasn't until at least 40 years after the death of the author, Herman Melville, that *Moby Dick* was acknowledged as a masterpiece. Sadly, Melville died virtually unrecognized.

MIGHTY SPERM WHALE *Moby Dick (below), a legend in American literature.*

HERMAN MELVILLE (1819–91)

Herman Melville (pictured below) was a prolific American novelist, who was famous for writing a series of books inspired by his early experiences as a whaler in the South Pacific.

He first set sail, in 1841, as a 23-year-old aboard the New Bedford sperm whaler *Acushnet*. The first few months aboard ship were so hard that, when they put into Nukahiva, in the Marquesas, Melville deserted. Within a few weeks he was heading for Tahiti, aboard the Australian whaler *Lucy Ann*. But that trip was also troubled and, after a rather disorganized mutiny, he was forced to spend a few days in a Tahitian jail. He was released quickly, though, and signed aboard his third ship in less than a year—the Nantucket whaler *Charles & Henry*—which took him to Lahaina, in Hawaii. By then he had had enough of whaling and eventually returned home as an ordinary seaman on the frigate *United States*.

When they were first published, Melville's books received a mixed reception from the public and critics alike. The first two, *Typee* (1846) and *Omoo* (1847), were extremely popular, but *Mardi* (1847) sold so few copies that it practically bankrupted its publishers. Later on, the titles *Redburn* (1849) and *White Jacket* (1850) helped to revive Melville's collapsing reputation, but his real masterpiece was *Moby Dick* (1851). Ironically, most reviews of *Moby Dick* were as savage as the starring whale itself and, to make matters worse, two years after publication, the unsold copies were destroyed by a fire. Herman Melville continued to write, but he could no longer survive by writing alone. Eventually, to support his family, he was forced to become a customs inspector in New York.

CHANGING ATTITUDES

Over the centuries, cetaceans have been viewed as
sea monsters, gods, guardians, reincarnations,
sources of food and income, and even living islands.

LIVING ISLAND *Tenth-century image*
of navigators camping on a whale's back.

The earliest known portrayal of a cetacean in art is an ancient drawing of an orca carved into coastal rocks in northern Norway, estimated to be some 9,000 years old. Several thousand years later, early Greek and Roman artists—inspired by the dolphin's apparent intelligence and kindness to humans—began to adopt dolphin motifs on vases, coins, mosaics, sculptures, and paintings. One of the earliest and better known of these is the dolphin fresco on the wall of the Queen's Room in the Minoan Palace of Knossos, on the island of Crete, which was painted by an unknown artist some 3,500 years ago.

The first cetologist in recorded history was the Greek philosopher and scientist Aristotle (384–322 BC). In *Historia Animalium*, he describes a variety of different species—dolphins, orcas, sperm whales, and right whales—and correctly classifies them all as mammals. He knew that cetaceans, unlike fish, breathe air; and he even made a distinction between the two major groups of whales, describing mysticetes as having "hairs that resemble hog bristles," instead of teeth.

EARLY BEGINNINGS
Greek philosopher
Aristotle (left).

The Roman scholar Pliny the Elder (AD 23–79) added to Aristotle's findings in his *Naturalis Historia*, which was a major source of scientific knowledge until the seventeenth century. Unfortunately, he did not discriminate fact from fiction, and included a confusing assortment of original firsthand observations with second- or thirdhand accounts, folklore, and superstition. However, he knew, for example, that some whales migrate to different places at different times of the year, and that they are sometimes attacked and killed by orcas.

MONSTER WHALES

A number of other ancient authors mentioned whales, and stories about them being as big as islands were especially popular in the Middle Ages, but it was several hundred years before they were really back in vogue. Records from the ninth century suggest that the Norwegians "prided themselves in knowing 23 species of whales." Later, in the thirteenth century, an anonymous account of ancient Iceland, called *Speculum Regale*, or *The Mirror of Royalty*, rather unfairly pronounced whales the only truly interesting sight the island nation had to offer.

For centuries, the seas and oceans were alive with imagined monsters—and none were thought more impressive

than whales. Their size alone filled people with terror, suggesting great strength and significance. Islamic text tells of a whale that was believed to support the world, and whales appear in Inuit myths about the beginning of the world. Meanwhile, the ancient Chinese believed that a strange mythical figure, with the body of a huge whale but the hands and feet of a human being, ruled the sea.

In Japanese mythology, a monster whale replaces the hare in Aesop's fable of the hare and the tortoise. After boasting that he is the greatest animal in the sea, the whale challenges a lowly sea slug to a race. The sea slug accepts and arranges for each of his sea slug friends to wait for the whale at different beaches along the route. On the day of the race, the whale surges into the lead almost immediately. He arrives at the first beach and calls out "Sea slug, sea slug, where are you?" The sea slug on duty calls back "What, whale? Have you only just arrived?" The baffled whale challenges the sea slug to another race, and the same

thing happens—the next sea slug is already waiting at the second beach. After several more impossible races, the whale finally admits defeat.

Myths and legends about whales began to disappear around the late fifteenth and early sixteenth centuries. With the rise of commercial whaling, people began to see whales more as potential sources of income than mythical monsters of the sea.

FRIENDLY DOLPHINS

Since the very first recorded contacts with dolphins, people have viewed them more favorably than whales. Dolphins have been depicted as gentle, trusting creatures, with an almost human intelligence. Stories of special friendships between dolphins and people have been around for at least 2,000 years. Dolphins have also been attributed with god-like qualities: some Australian Aborigines, Polynesians, and Native Americans, for example, traditionally regard them as spirits or messengers of the gods.

The ancient Greeks held dolphins in such high esteem that killing them was tantamount to killing a person, and both crimes were punishable by death. The Roman world was captivated by their extraordinary friendliness and alleged passion for music. But

ANCIENT AND MODERN *The dolphin fresco in the ancient Minoan Palace of Knossos (above). A group of whale watchers experiences friendly gray whales firsthand (below).*

the Romans were less emotive and sentimental about nature than the Greeks, and their dolphin stories tended to have decidedly tragic endings.

Many of the ancient dolphin myths contained an important truth or moral— in particular, they taught the value of respecting the natural world, and warned that anyone who treated it with contempt was sure to experience the wrath of the gods.

MODERN ATTITUDES

In some ways, modern attitudes toward cetaceans are probably more complex than ever before. Whales, dolphins, and porpoises tend to mean different things to different people: for example, they are viewed as a nuisance by some fishers, as an exploitable

resource by the owners of marine parks and by the dwindling commercial whaling industry, and as a source of endless fascination and wonder by millions of whale watchers around the world.

In general terms, our attitude toward dolphins has changed surprisingly little. Today, a great many people share the same feelings as our ancestors. But our general attitude toward whales has changed—particularly over the last 20 or 30 years. For most people, they are no longer the terrifying monsters they once were, or merely animals to be killed for profit. At last, albeit gradually, we are learning to respect both whales and dolphins, as extraordinary animals that deserve special attention.

PERSONAL ENCOUNTERS

The striking level of trust that many whales, dolphins, and porpoises show toward people is deeply moving.

We hunt them without pity, pollute their homes, drown them in fishing nets, and catch all their fish—yet they seem so ready to accept us as friends. Some actively seek out human company. Indeed, they are frequently so inquisitive that whale watchers often find themselves face-to-face with whales or dolphins, wondering who is supposed to be watching whom. Some gray whales in their breeding lagoons on the Pacific coast of Baja California, Mexico, even lift their heads out of the water and wait to be petted by incredulous tourists.

"Friendly" dolphins are particularly intriguing. One of the first to become an international celebrity was a Risso's dolphin named Pelorus Jack. For about 24 years, from 1888 until 1912, he escorted ferry steamers through the narrow French Pass, on the northern tip of the South Island, New Zealand. Pelorus Jack met an untimely death in April 1912, when he was harpooned by the crew of a Norwegian whaling boat anchored offshore. He is still fondly remembered to this day.

At least 10 dolphin species are known to have shown a liking for human company, although the bottlenose is easily the friendliest of them all. One of the most famous is Fungie, known to many simply as "the Dingle dolphin." Fungie appeared in Dingle Harbour, County Kerry, Ireland, in the mid-1980s, and has stayed ever since, becoming a great tourist attraction in Ireland (see also p. 238).

DOLPHIN RESCUE

Stories of dolphins helping human swimmers in distress or protecting them from sharks are quite common. In November 1988, two sailors were shipwrecked in rough seas off the coast of Indonesia. When their ship disappeared beneath the ocean waves, a group of dolphins appeared. The animals nudged and guided the men throughout the night, until they finally reached the safety of dry land.

Some years ago, a woman was in the water with a female humpback whale and her calf off the coast of Tonga, in the South Pacific, when a large bronze whaler shark suddenly appeared. Within moments, the woman was surrounded by spinner dolphins who stayed with her until the shark had lost interest and swum away.

More recently, in July 1996, a man jumped into the Red Sea to swim with a group of five bottlenose dolphins. Within minutes he was being attacked by a shark—later thought to be a tiger shark—which lifted him nearly 5 feet (1.5 m) out of the water. While one of his friends

A CLOSE ENCOUNTER

We were at the point where Chatham Strait meets Frederick Sound, in southeast Alaska. A region of remote, rugged wilderness. Majestic snow-capped mountains rise steeply from sea level to form sheer-sided fiords; towering glaciers calve into freezing cold seas, and waterfalls cascade through dark green valleys of ancient coniferous forests.

We were in Alaska to see humpback whales doing the unimaginable: fishing with nets made of bubbles. In fact, we were about to witness one of the greatest wildlife spectacles on Earth, although it was almost impossible to predict exactly when or where it was likely to happen.

TRUE FRIENDS *Bottlenose dolphins (left) are well known for enjoying human company. Another "friendly" cetacean is the gray whale (above left), often seen welcoming whale watchers in boats.*

HEALING POWER *Bottlenose dolphins are believed to provide therapeutic benefits for young children in need (right).*

jumped into an inflatable boat and raced to the rescue, three of the dolphins surrounded the man, slapping their flukes and fins on the surface of the water, preventing a further attack until he could be lifted out of danger.

DOLPHIN THERAPY

Research in several countries suggests that dolphins may be able to trigger the healing process in people. It is a claim that has had more than its fair share of sceptics, and is especially difficult to demonstrate scientifically. However, there are so many accounts of dolphins alleviating cases of chronic depression or anxiety, enhancing recovery from life-threatening illnesses such as cancer, and even speeding up the learning potential of children with disabilities, that biologists and doctors around the world are beginning to take these claims seriously.

Horace Dobbs, a specialist in the healing power of dolphins, tells the story of a man who was being treated for chronic depression and had not been able to work for 12 years. In his own words, he was living in a "black pit of despair." Then, in 1985, Dobbs took the man to swim with Simo, a friendly bottlenose dolphin living near the tiny fishing village of Solva, in Wales. Within moments, the two became inseparable. The man broke his silence and spoke to Simo as if he were talking to an old friend. For the first time in all those years, the black cloud had started to lift.

Suddenly, a bubble broke the surface, no more than 30 feet (9 m) from where we were standing. It was roughly the size of a dinner plate and was quickly followed by another bubble beside it, and then another, and another. Within a few moments, a huge circle of bubbles had formed on the water's surface.

We had a split second to prepare ourselves. Then it happened. A mind-boggling 14 humpback whales suddenly erupted from the water in one great foaming mass. About 450 tons (400 tonnes) of gaping mouths (shown above) simultaneously exploded—literally within yards of where we were leaning

against the railings of the boat. With water gushing down extended throat pleats, and clouds of leaping herring, the whales rose to a height of nearly 20 feet (6 m) before sinking back into the water. The wash rocked us from side to side.

By the time the water had settled, and we were beginning to calm down, it was as if nothing had happened. Like a dream. We were left with just a loon calling in the distance. The whales had simply disappeared, leaving behind no trace of their spectacular performance.

Mark Carwardine,
extract from diary, July 1995

HOW LITTLE WE KNOW

The lives of most whales, dolphins, and porpoises are still shrouded in mystery.

Despite a flow of new information, we know surprisingly little about the majority of cetacean species and our knowledge of the others is frustratingly patchy. We have learned enough to start asking the right questions, yet the deeper we delve, the more we realize there is still to learn.

The problem is that these are among the most difficult animals in the world to study. They often live in remote areas far out to sea, spending most of their lives under-water, and showing little when they rise to the surface to breathe. No wonder, as one researcher commented at a recent conference, that our knowledge of many species has only progressed from almost nothing to just a little.

DEAD ANIMALS

In the early days of cetacean research, the little information we had came mainly from dead animals washed ashore or killed by whalers and fishers. Even today, profes-sional post-mortems can teach us a great deal about poorly known species, such as beaked whales. A single cetacean car-cass can provide an amazing amount of data that, when combined with other data, becomes invaluable.

The precise details of a post-mortem depend largely on the aims and objectives of the study, and the information required. Typically, it consists

STRANDING *of long-finned pilot whales pro-vides an opportunity for a post-mortem (right). A baby dolphin is examined in the laboratory (below).*

of three distinct stages, often carried out by different people with a diverse range of skills. The first stage takes place on site (if the animal cannot be easily transported) or in a laboratory, and involves a general overview: everything from taking external body measurements and noting its general physical condition to identifying its sex and check-ing its reproductive status.

The second stage, normally in a laboratory, involves a much more detailed analysis: estimating the age of the animal, checking key body tissues for pollutants, identi-fying its stomach contents, examining genetic material, and much more. The final stage can often be the most

time-consuming, and usually takes place in an office in front of a computer screen. This is the analysis of all the results. What do they mean? How do they compare with findings of similar studies? Are they statistically significant? What further research needs to be done? These and other questions have to be answered before, finally, the work can be written up and published in a scientific journal.

CAPTIVE ANIMALS

In the mid-1800s, another source of information became available. Since then, biologists have been able to study cap-tive animals—usually dolphins and other small species that can be confined in a concrete tank—with varying results.

On moral grounds, many people would argue against keeping cetaceans in captivity for research purposes (or for any other reason). But there are also concerns about the research itself. Inevitably, an animal in captivity will behave abnormally—at least part of the time—and this can give a thoroughly distorted picture of its behavior under natural conditions. Even if it is in a large enclosure, it will be unable to carry out all the normal activities of its wild relatives. At the very least, conclusions drawn from research on captive animals have to be treated with caution. Many studies can be carried out satisfactorily only under natural conditions.

But there are advantages in studying cetaceans under controlled conditions. They can be observed at close range for 24 hours a day. Knowing their age, sex, reproductive status, and level of dominance is another major benefit enabling researchers to study their biology and behavior in the most minute detail. It is also useful to be able to undertake specific scientific experiments, under tightly controlled conditions, on everything from sleep requirements to the physiology of diving.

GOING OUT TO SEA
A great deal of information can be obtained only by studying the animals, wild and

Knowledge is the true organ of sight, not the eyes.

Panchatantra,
ANON (c. fifth century),
(trans. by FRANKLIN EDGERTON)

free, in their natural environment. Few people took up this enormous challenge until the late 1960s and early 1970s, and it is only in the past decade or so that research into wild cetaceans really developed into the sophisticated and popular branch of natural science we know today.

Early pioneers focused mainly on counting whales, dolphins, and porpoises at sea. Observers were posted at lookouts on land, in boats, or in light aircraft, and used relatively simple methods to calculate population and group sizes. These days, there is still no perfect way to count cetaceans, but the techniques are being refined all the time and modern surveys take into account everything from the number of animals likely to

have been missed along a cruise track to varying sea and weather conditions. At the same time, research has gone well beyond mere population estimates to social and behavioral studies, combined with careful note-keeping, resulting in many exciting discoveries in recent years.

PATIENCE
Despite the difficulties, it is an exciting time to be involved in wild whale, dolphin, and porpoise research. We are just beginning to understand the intricacies of their natural behavior, diving capabilities, social organization, feeding techniques, and many other aspects of their daily lives. In fact, in recent years, growth in our knowledge has been nothing short of remarkable.

The main requirement for any study of them is patience. It can take many years to collate all the tiny snippets of information necessary to assemble a single coherent picture. It is something like putting together an enormously complicated jigsaw puzzle, one piece at a time, in which every individual piece raises new questions. The main difference, of course, is that the complex cetacean puzzle will never be completely finished.

GATHERING DATA *A captive bottlenose dolphin (above) is studied by biologists. Feces collected in the Sea of Cortez (right) will be examined later in a laboratory to reveal what blue whales are eating.*

HIGH-TECH RESEARCH

In the past decade, the sophistication and variety of information-gathering techniques for whale research in the wild have grown beyond our wildest dreams.

Today's whale researchers spend long hours patiently observing the same animals day after day, month after month, and year after year. They also make do with some fairly basic equipment: it is not uncommon to see them using children's fishing nets to scoop up prey samples, or donning goggles, holding their noses, and hanging their heads over the side of a boat to see what is going on underwater.

But at the same time, researchers frequently enlist the help of state-of-the-art equipment and space-age research techniques: satellites in space, radio transmitters, deep-sea submersibles, high-tech directional hydrophones, complex computer programs, fiber optics, deep-water video probes, DNA fingerprinting, and, most recently, the US Navy's submarine tracking system is now part of the modern whale researcher's armory.

SATELLITE TELEMETRY

Unlikely as it may sound, the answers to many of the more perplexing questions about whales probably lie in space. To be more precise, they lie in satellites in space. The idea of tracking whales by satellite may seem far-fetched, but it has already been tried and tested on several species with considerable success.

Researchers normally use a modified gun or crossbow to

EYE IN THE SKY *The European remote-sensing satellite, seen over the Netherlands (above), can track tagged whales wherever they go. The small transmitter attached to a beluga (right) collects an amazing amount of data.*

fire a small, battery-powered transmitter into the thick blubber of the whale's back. This beams signals up to orbiting communications satellites, which relay the coded information back to receiving stations on Earth. Since satellites continually scan the entire surface of the Earth, the animals being studied can be tracked wherever they go. The information is then sent to researchers sitting in front of computers in offices. These days, in addition to signaling a whale's movements, satellite transmitters can provide a

great deal more information, such as swimming patterns, heart rate and dive depths, water temperature and pressure, and much more.

Satellite tracking is not without its critics. It is an invasive research technique, perhaps causing at least some temporary discomfort to the whales being studied. It may even affect their behavior—and, therefore, the research results. But the transmitters themselves are continually being refined and, with considerable advances in microelectronics, each year they get smaller and smaller.

TAGGING *A small harpoon is used to implant a probe into the skin of a blue whale (above). Molecular model (right) of the famous DNA double helix.*

The latest models are just a few inches long, yet they can gather information that was previously either difficult or impossible to obtain.

LISTENING FOR WHALES

Our ability to learn about whales, simply by listening to the sounds they make, is no longer science fiction but real scientific fact. An astonishing amount of information can be obtained with hydrophones, sophisticated underwater microphones (see p. 115). This field is undoubtedly a challenging area of research—in some ways, it's rather like trying to find out what goes on in New York merely by dangling a microphone from the top of the Empire State Building—but it is already beginning to revolutionize our knowledge about whales.

A major breakthrough came in 1992, when the US Navy decided to give some researchers access to its so-called Integrated Underwater Surveillance System. This consists of a series of sophisticated underwater listening stations on the Atlantic sea floor, which, in effect, turn the ocean into a huge sound studio. It was originally designed for tracking Soviet submarines by listening to their faint acoustic signals, but it has tremendous potential for whale research. Already, thousands of whales have been heard calling, and the experts are beginning to track their movements.

It may even be easier to count whales by listening for them rather than by the age-old technique of looking for them. Researchers at Cornell University in Ithaca, New York, have developed a sound-sensitive "acoustic telescope" that seems to be better at spotting whales than human whale watchers. The "telescope" consists of 16 hydrophones on a line about a mile long, which is towed behind the research vessel. By listening carefully to the sounds the whales make, it is possible to identify the species, to distinguish individual animals, and to work out what they are doing. One of its major advantages, of course, is that it can evaluate what is going on underwater—whereas a whale watcher standing on the deck of a boat, with just a pair of binoculars, can see only what is going on at the surface of the water.

DNA FINGERPRINTING

Do different calves with the same mother have the same father? Are fathers involved in their upbringing? Are individuals that spend a lot of time together related? These and other intriguing questions can be answered just by examining a small piece of a whale's skin. It is the genetic material, or DNA (deoxyribonucleic acid), that is so revealing.

The DNA itself is a type of instruction manual for the organization and assembly of the body's proteins. All body cells contain an exact replica of this manual, and almost all the "pages" go toward making the animal what it is: a harbor porpoise, a striped dolphin, a humpback whale, and so on. The few pages left help to differentiate between individuals of the same species. Just as one person's fingerprints differ from everyone else's, no two animals have exactly the same DNA, yet related animals show some similarities. The clever detective work involved in interpreting this highly complex information is called "DNA fingerprinting," and it is already proving invaluable for whale research in the wild.

DNA fingerprinting is somewhat controversial, because it normally involves firing a small dart from a crossbow to take the tiny sample of skin essential for genetic analysis. So some biologists are beginning to collect tiny pieces of skin that flake off naturally, like whale dandruff, as the animals dive, instead of taking samples.

Accuse not Nature; she hath done her part;
Do thou but thine.

Paradise Lost,
JOHN MILTON (1608–74), English poet

CHAPTER TWO

SAVE *the* WHALES

THE HEYDAY of WHALING

Since its early beginnings nearly 1,000 years ago, commercial whaling has had a long and checkered history.

MORE RECENT TIMES *Several whales are being weighed (above) at a railway station in Durban, South Africa in 1939.*

Its origins are unrecorded, but the first large-scale commercial whaling was probably by Basques from the coasts of France and Spain. By 1200, they were hunting right whales in the Bay of Biscay, and had found a market for nearly every part of the whale carcass—even the excrement was used as a red fabric dye.

By the beginning of the seventeenth century, Britain, Holland, and other nations had realized whaling's potential and were busy competing for control of the richest grounds. The American whaling industry began soon after, in the 1640s, when early colonists first ventured from the green shores of New England to hunt right whales.

WHALING SCENE
A nineteenth-century whaler flying the American flag.

In many ways, the next 200 years were like the gold rushes that later swept across western North America. Like prospectors searching for big strikes, whalers moved relentlessly from one hunting ground to another, killing all the animals in one place, then moving on to the next. Whaling rapidly developed into a major industry and, on the way, exposed new worlds to the mixed blessing of Western civilization.

THE AMERICAN WHALER
Everyone involved in a whaling expedition was taking a risk. Ship owners gambled many thousands of dollars to outfit their vessels. While some ships came back with their holds full again and again—making their owners exceedingly rich—others left on their maiden voyages and vanished without trace—causing the owners to lose their investment completely.

Officers and crew, on the other hand, gambled not with money, but with their lives. Many never made it home from their long and hazardous journeys. If they didn't become the victims of violent storms, rogue whales, reefs, accidents, disease, or fighting, they often jumped ship.

An American whaling vessel typically left port with a crew of about 30–35 men. These included the captain, 4 harpooners, 4 mates, a cooper (for making and repairing barrels), a steward, a blacksmith, a cook, a cabin boy, and

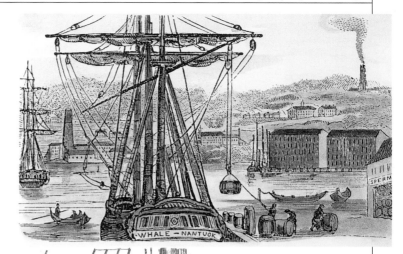

ARTISTS' IMPRESSIONS *An 1852 engraving (right) of a whaling ship at a New England port. An undated print (below) shows a seaman being flogged.*

15–20 ordinary seamen. While the captain and his mates lived in fairly comfortable accommodation toward the stern of the vessel, the ordinary seamen lived together in cramped, dark, filthy conditions near the bow.

Food was a constant source of complaint. Hardly anyone ate whale meat because the taste was considered too strong, but the alternatives were little better. Occasionally, they would be able to trade for fresh meat, fruit, and vegetables, but the basic diet was salted pork or beef, with thick, hard crackers, and coffee or tea. Floating the crackers in a hot drink, until the maggots crawled out and could be skimmed off, was a common practice.

It was also hard for wives and girlfriends left behind. Many whaling voyages lasted for four or five years, and there were no guarantees the men would ever return. Communications were almost non-existent, despite a system of makeshift "post offices" in some parts of the world. In the Pacific, for example, ships heading out to the whaling grounds would leave letters on the Galápagos Islands in a wooden box covered by an enormous tortoise shell, ready to be picked up by other ships heading home. But it was a rather hit-and-miss affair—one woman wrote more than a hundred letters to her husband during his voyage, yet he received only six.

As word filtered out from the ports of New England,

home to the American whaling armada, the romance of whaling was soon outweighed by real-life stories of danger, cramped and filthy conditions, tyrannical officers, boredom, and the reality of long absences from home. A popular saying at the time was "One voyage on a whale ship is one too many." One log keeper on the Yankee whale ship *Brewster* commented: "My opinion is that any man who has a log hut on land with a corn cake at the fire and would consent to leave them to come … on a whale voyage is a proper subject for a lunatic asylum."

With so much bad publicity, combined with the rapid expansion of whaling after 1815, New England seaports were soon finding it hard to recruit competent crews. It was even harder by the mid-1800s, when the industry had to compete with the lure of free land and gold in the

West. Unscrupulous "shipping agents" were often hired to assemble ordinary seamen—by whatever means necessary. They rounded up naive young men from up and down the eastern seaboard, and across the Midwest, tempting them with false promises of money and adventure. Worse still, it was not an uncommon event for men to wake up after a night on the town with a massive hangover in a rock-hard bunk on their way to whaling grounds hundreds or thousands of miles from home.

In an age before petroleum or plastics, there were huge amounts of money to be made from whaling—but they rarely filtered down to the crew. Everyone received a percentage of the net value of the cargo: the ship's owner received about 66 percent; the captain up to 10 percent; the mates, harpooners, and coopers up to 1 percent each; the ordinary seamen about 0.6 percent; and the cabin boys no more than 0.4 percent. Even then, a variety of "expenses" were deducted from these meager wages and, ultimately, the lowly crew earned considerably less than unskilled workers on shore.

USES *of* WHALES, DOLPHINS, *and* PORPOISES

Whales once provided valuable raw materials for thousands of everyday products, but nowadays more acceptable substitutes are used.

Somehow, the whaling industry managed to find money-spinning uses for nearly every part of a whale: from its oil, meat, and baleen, to its skin, blood, and tendons. The sheer variety of products was quite astonishing and, for several centuries, they touched almost every aspect of daily life in Europe and North America.

WHALE OIL

The thick layer of blubber around a whale is composed of a fibrous, fatty material, honeycombed with large cells, each filled with oil. Extracted simply by heating the blubber to a very high temperature, whale oil was known as the "liquid gold" of the whaling industry.

Over the years, it has been used to make a bewildering range of products, including soap, shampoo, detergent, lipstick, margarine, cooking fats, ice cream, crayons, paints, polishes, linoleum, machine lubricants, and even glycerine for explosives. In particular, the homes of Europe and North America were lit by whale-oil candles and lamps: Newton, Voltaire, Swift, and Bach would all have worked by the light of whale-oil lamps.

Sperm whales produced the finest and most valuable oil, not from their blubber, but from inside their huge, cavernous heads. Known as spermaceti oil, it was initially used to make high-quality candles but later became a sought-after lubricant. Some people even attributed medicinal properties to it, claiming excellent results for anything from renal colic and catarrh to open wounds and pulmonary ulcers.

BALEEN PLATES

Baleen, or whalebone as it is misleadingly known, was used in a variety of products. Whip handles, riding crops, shoehorns, umbrella ribs, bristles for brushes, watch springs, store shutters, fans, fishing rods, and even tea trays were all made—or strengthened—with this unlikely raw material.

PASSING FASHIONS *Sperm whale teeth carved by whalers in idle hours are now valued collectibles known as scrimshaw (top). Hourglass figures (above) by courtesy of whalebone corsets. The midnight oil Voltaire (left) and others burned came from whales.*

It also had quite an influence on nineteenth-century fashion. Women squeezed their bodies into elaborate corsets stiffened with baleen, or wore baleen hoops under their skirts. It was even used to maintain the curls in fancy wigs.

WHALE MEAT

Whale meat has never been in great demand in Europe or North America, except as a necessity during times of war. Some was ground up into fertilizer or used as animal fodder and in canned dog and cat food but, for many years, much of it was thrown back into the sea.

Norwegians, Icelanders, Koreans, and a few other nationalities do eat whale meat, but the main commercial market has always been Japan. In the past, Japanese whalers profited from this healthy demand in their own country but, these days, whale meat is sold primarily as a delicacy in smart Japanese restaurants.

NOTHING TO WASTE

Whale skeletons and bones have been put to some particularly strange uses over the years. The Vikings made chairs from whale vertebrae, for example, while Faeroe Islanders used pilot whale skulls to build their farm walls. Meanwhile, the whalers themselves frequently carved and decorated whale bones and teeth, making buttons, chess pieces, snuffboxes, cufflinks, necklaces, and a variety of other fancy articles.

Whale skin was used to make bootlaces, or turned into bicycle saddles, handbags, and shoes. The blood became

AMBERGRIS: WHALE GOLD

Ambergris was once regarded as the most valuable of whale products. It could be obtained without encountering the whale itself. Ambergris occurs as a waxy, solidified lump occasionally found floating in the sea, or washed ashore. It is formed in the lower intestines of some sperm whales, and disgorged when the whales vomit. Despite its unpleasant origin, "a little whiff of it," according to the British poet Alexander Pope, "… is very agreeable." Primarily made of ambrein, a fatty substance similar to cholesterol, it hardens on exposure to air, and often contains squid beaks.

Whenever whalers found a lump of ambergris in the sea, or removed it from a recently killed sperm whale, they whooped and cheered. It was worth its weight in gold. Originally used as a medicine to treat indigestion, convulsions, and a variety of other ailments, and also as an aphrodisiac, it became invaluable later as a fixative in perfumes and for making cosmetics.

Since the early 1900s, some huge lumps of ambergris have been found, a few of them looking more like giant rocks or boulders than something that could have come out of a sperm whale. The largest ever recorded, weighing an incredible 1,400 pounds (635 kg), was found near New Zealand by a British whaler.

WHALEBONE CASKET

This beautiful Frankish object (left) is decorated with an Anglo-Saxon runic inscription.

There is no such thing as absolute value in this world. You can only estimate what a thing is worth to you.

"Sixteenth Week,"
My Summer in a Garden,
CHARLES DUDLEY WARNER
(1829–1900), American writer

an ingredient in sausages, and was used in fertilizers and adhesives. Even the tendons were turned into strings for tennis rackets and "catgut" for surgical thread. The liver was a source of vitamin A, while the connective tissue yielded a gelatin for use in sweets and photographic film. Perhaps most unlikely of all, in parts of Scandinavia the intestines were used as a substitute for window glass.

One by one, all of these whale products have gradually disappeared. Petroleum, plastics, and other more acceptable substitutes replaced some of them, and changing fashions reduced the demand for others. Also, whale numbers rapidly declined, so the whaling industry had fewer raw materials and, ultimately, no products to sell.

MODERN COMMERCIAL WHALING

A series of technological advances in whaling vessels, killing equipment, and hunting methods spelled disaster for the world's whales.

Modern commercial whaling really began in the mid-nineteenth century, when the Norwegian whaler Svend Foyn developed an explosive harpoon that could be fired from a bow-mounted cannon. Foyn's invention was the first in a remarkable series of developments that revolutionized the whaling industry over a period of about 60 years, and greatly increased the number of whales being killed.

Other major advances included faster and more maneuverable catcher boats; longer and stronger harpoon lines; better winches to relieve some of the strain as the whales struggled for their

ABUNDANCE *The aquatint (above) of a painting published in 1825 shows a South Sea whale fishery in the early days.*

lives; machinery for inflating the carcasses with air, to make them float; and enormous factory ships that could process the whales far out to sea. New and improved techniques for processing whale products were also developed and, meanwhile,

navigational aids improved, weather forecasting became more accurate, and new technology helped to make whaling much safer.

SVEND FOYN 1809–94

Norwegian Svend Foyn (left) has been dubbed the inventor of modern whaling.

Indirectly, he is probably responsible for the deaths of more whales than anyone else in the business.

Born in 1809, he grew up in a small harbor town called Tønsberg, near the capital, Oslo. From an early age, he was determined to be a seaman, even though his father had died at sea when he was just four years old. He embarked on a

career in the coastal freight business, but left that in 1846, and turned his attentions to sealing. Hunting seals around Jan Mayen, a tiny Arctic island between Iceland and Spitzbergen, quickly made him rich. But when the profits began to decline in the early 1860s, he began to look for new challenges.

He had noticed large numbers of whales, mainly fin and blue, on his trips to and from the sealing grounds. Realizing that they had somehow escaped the attentions of whalers, he decided to have a go at hunting them himself.

A deeply religious man, he later wrote that "God had let the whale inhabit [these waters] for the benefit and blessing of mankind and, consequently, I considered it my vocation to promote these fisheries."

During the years that followed, Foyn transformed the whaling industry by building the first steam-driven catcher ship and inventing an explosive harpoon that could be fired from a fixed cannon.

He was active to the end of his life and died, as an 85-year-old, while sponsoring a whaling expedition to Antarctica.

KEY TECHNOLOGICAL DEVELOPMENTS IN MODERN WHALING

Until just over a century ago, the technique of whaling had barely changed for more than 700 years. Whalers chased their quarry in open rowing boats, and normally killed them with hand-held harpoons, just as the Basques had done when they hunted right whales in the Bay of Biscay during the eleventh century. But in the 1850s, the era of modern commercial whaling began with a vengeance—and the whalers have not looked back since.

1852 The first successful explosive harpoon (above) was invented by an armorer from Connecticut, United States. Known as the bomb-lance, it consisted of an explosive missile armed with a time-delay fuse that killed or mortally wounded the whales—normally at a safe distance from the whalers.

1857 The first auxiliary steam engines were installed in British whaling ships and, in the years that followed, many other whalers installed similar engines in their vessels.

1859 The first purpose-built steam whaling ships were successfully launched. The extra power gave them increased speed and safety, although once on the whaling grounds, the whalers continued to hunt from traditional rowing boats.

1863 Norwegian sea captain Svend Foyn built a new, 82-foot (25 m) steam-driven schooner called the *Spes et Fides* (meaning "Hope and Faith"). The first of the modern whale catchers, this combined the functions of the open rowing boat and the ship in a single vessel. It was relatively easy to maneuver and yet fast enough to chase the whales by itself. (A similar vessel, from the 1890s, is seen above.) No species of whale, no matter how fast a swimmer, was now safe.

1865 A whaling captain from New Bedford, in the United States, invented the darting gun, a kind of explosive harpoon with the accuracy of a conventional harpoon.

1868 After several seasons of experimentation, Svend Foyn developed an explosive harpoon that could be fired from a cannon mounted on a swivel in the bow of a whaling vessel (shown below, manned by skipper Duncan Greys). The cannon was solidly made to absorb the recoil and so well balanced that it could be aimed with great accuracy. Screwed to the tip of the harpoon was a grenade, consisting of a detonator and a sack of black gunpowder in a steel container equipped with barbs. The barbs were designed to open on impact, ensuring that the harpoon was firmly embedded in the animal's flesh, and the grenade exploded inside its body two or three seconds later. This awful weapon precipitated an enormous increase in whaling worldwide and is still in use today, albeit in modified form.

1925 The growth of the whaling industry was limited by a need to return constantly to shore-based stations to process the whales. The final dramatic development, floating factory ships (such as this one below, whaling in South Georgia), solved this problem. The new ships were designed to accompany fleets of catcher boats far out to sea and, with stern slipways to winch the whales on board, gave a new level of efficiency to the hunt. It was in 1925 that the first factory ship arrived on the Antarctic whaling grounds.

THE AFTERMATH *of* WHALING

*In the space of a few hundred years,
untold millions of great whales were killed around the world.*

The final death toll is shocking: about a million sperm whales, at least half a million fin whales, more than 350,000 blue whales, nearly a quarter of a million humpbacks, and literally hundreds of thousands of others. Two million whales were killed in the Southern Ocean alone. Some years were particularly bad: in the 1930–31 season, 28,325 blue whales were killed; more than 30 years later, in 1963–64, no fewer than 29,255 sperm whales met their death—we still had not learned the lessons of the past. Today, we are left with merely the tattered remnants: in most cases, no more than 5–10 percent of the original great whale populations remain.

By the time the animals were given official protection it was almost too late and, indeed, some species may never recover. The rarest whale in the world, the northern right whale, came so close to extinction that, even after more than 60 years of official protection, its population is represented by only about 300 survivors. Meanwhile, the bowhead whale has all but disappeared from vast areas of its former range.

But there is some good news, as well. Against all the odds, several species appear to be bouncing back. The gray whale (top) is the ultimate success story. The North Atlantic stock was probably wiped out by early whalers, and only a remnant population survives in the western North Pacific. But in the eastern North Pacific, it has made such a dramatic recovery that it is now believed to be at least as abundant as in the days before whaling. Meanwhile, there has been a dramatic increase in blue whale numbers off the coast of California, where about 2,000 animals gather for the summer and autumn (the largest animals on Earth have shown little sign of recovery elsewhere). Humpbacks and southern right whales are also making a comeback in many places.

NO WHALES: NO WHALING

Different whales have been in vogue at different times during the history of commercial whaling. Their popularity depended on the market value of their products, the whaling techniques available at the time, and the availability of the species (which, naturally, declined as whaling intensified).

Right whales were the first targets for commercial whalers and have been hunted for hundreds of years, since the eleventh century. Named for being the "right" whales to hunt, they were easy to approach, swam slowly, lived close to shore, had a tendency to float when dead, and provided large quantities of valuable oil, meat, and baleen. As their numbers declined, and whalers ventured farther

FAST SWIMMER *The blue whale (left) was too difficult to catch until the introduction of explosive harpoons and steam-powered vessels.*

Whom man kills, him

God restoreth to life.

"Fantine," *Les Misérables,*
VICTOR HUGO
(1802–85), French novelist

out to sea, sperm whales and other species became the new focus of attention. The swift and powerful blue whale was safe from whaling until the late 1800s, when explosive harpoons and steam-driven vessels became widely available, but it soon became the most sought-after species. Then, as blue whale stocks were depleted, whalers shifted their attentions to fin whales. One by one, the great whales were hunted almost to the point of extinction. The minke whale became sought after only when its larger relatives could no longer support the whaling industry; it is now the only whale being hunted commercially and within the confines of international law.

DYING INDUSTRY

As the number of whales declined, the whaling industry itself was threatened with extinction. But it was not until the twentieth century that whalers realized that their own survival depended on an abundance of available whales. Eventually, they joined forces and adopted the 1931 Convention for the Regulation of Whaling, which began, albeit rather inadequately, to regulate the industry. The International Whaling Commission (IWC) was created in 1946, and has regulated whaling ever since.

WORLDWIDE PROTECTION

The first whale to be given worldwide protection from commercial whaling was the bowhead, in the early 1930s. Most other species have been afforded regional or worldwide protection in the years since. This culminated, in 1986, with a moratorium on commercial whaling which, finally, granted worldwide protection for all the great whales.

Unfortunately, the killing often continued long after official protection was declared (and, in some cases, continues still today) but the following dates still have considerable significance:

1931	Bowhead whale (became US law in 1935)
1935	Northern right whale
	Southern right whale
1946	Gray whale (pictured right)
1966	Humpback whale
	Blue whale
1979	Sei whale (except in the Denmark Strait west of Iceland)
1984	Sperm whale
1986	IWC moratorium takes effect

PROTECTED *The humpback (left) and southern right whale (below) populations are recovering since the introduction of whaling bans.*

WHALING IS CRUEL

Despite all the advances made in the whaling industry, it is still virtually impossible to kill a whale humanely. It is extremely difficult to hit the vital organs of a moving animal—from a moving vessel—regardless of the accuracy of the harpoon gun.

Most whales are killed within a few minutes—which is bad enough—but some struggle in agony for much longer. In 1993, for example, a minke whale took 55 minutes to die a slow, lingering death at the hands of Norwegian whalers. It is hard to imagine the pain and suffering it experienced after an explosive harpoon had blown a huge, gaping hole in the side of its body. Japanese whalers use electric lances to kill harpooned whales, in a vain attempt to speed up the process. But the electric charge is grossly insufficient to induce death and, instead, causes violent muscle contractions and probably even more stress and suffering.

As one ex-whaler commented: "If whales could scream, whaling would have stopped many years ago."

THE INTERNATIONAL WHALING COMMISSION

Since 1946, government representatives from around the world have met every year, under the auspices of the IWC, to discuss whaling.

Originally, the IWC was established to make possible the orderly development of the whaling industry. It actively encouraged the development of whaling and, as a result, more than two million whales were killed during the organization's first 30 years—despite the advice of scientists on the Scientific Committee.

More recently, the IWC has been working toward better protection of cetaceans But some of its 39 member countries are still strongly pro-whaling. Consequently, IWC debates are frequently passionate and heated, and often involve behind-the-scenes politicking, bribery, and frequent threats.

WHALE PEACE TREATY

In 1982, there was a major breakthrough when IWC members voted for an indefinite moratorium on commercial whaling. But this world peace treaty for the whales left too many loopholes to work

COMMERCIAL AND SCIENTIFIC WHALE CATCHES UNDER THE IWC

Country	Region	Species	1994	1995	1996
Norway	Northeast Atlantic	Minke whale	279	217	388
Japan	North Pacific	Minke whale	21	100	77
Japan	Antarctica	Minke whale	330	440	440
Total:			630	757	905

effectively. Since then, an incredible 57,391 whales have been killed, including minke, fin, sei, Bryde's, humpback, gray, sperm, and bowhead whales. Even since 1985–86, when the whaling ban actually came into effect, Japan, Norway, Iceland, Russia, Korea, and indigenous whalers in several other countries, have killed some 21,760 whales between them.

In 1994, members of the IWC approved the Southern Ocean Whale Sanctuary, covering an area of some 20 million square miles (50 million sq km) around Antarctica. The sanctuary was designed to protect a critically important feeding ground for seven species of great whale. But even this is not sacrosanct, since Japan continues to

hunt minke whales within the "protected" area.

Two countries, Norway and Japan, continue to hunt whales in blatant defiance of world opinion and the intentions of the IWC. Far from bringing their whaling activities to an end, they are steadily expanding them.

Norway officially objected to the moratorium and so, under IWC rules, is legally entitled to continue hunting whales. In the first 11 years after the ban, its industrial whaling fleet killed 2,011 minke whales. Meanwhile, its annual quota, set by the Norwegian government itself, is steadily increasing; and its whaling fleet of several dozen vessels spreads farther out across the northeast Atlantic.

Initially, Japan took the same course of action, but then withdrew its objection and renamed its whaling "scientific research." This takes advantage of another

DEAF EARS *The IWC meets (top) in Brighton, Britain, but Norwegian whalers continue to operate in spite of bans. A minke whale (left) is winched aboard.*

THE GERLACHE STRAIT *(above) is part of the vast Southern Ocean Whale Sanctuary in Antarctica.*

major loophole: the IWC allows any number of whales to be taken "for purposes of scientific research." A limited amount of research is undertaken, yet no new knowledge is being gained, and the whale carcasses are processed commercially. In the same 11-year period since the moratorium, Japan has killed 6,083 minke, 634 Bryde's, and 400 sperm whales.

HISTORY OF ABUSE

Many whaling nations have a checkered history of undermining, ignoring, and abusing the international regulations established to protect whales. For example, in the early 1990s, it was discovered that Russia had been falsifying its whaling data for more than 30 years. As well as killing protected species, it had grossly exceeded its IWC quotas for other species. In April 1996, 6½ tons (6 tonnes) of Norwegian whale meat was confiscated by Japanese customs officials. Disguised in a mackerel shipment, it was the first instalment of a 66-ton (60-tonne) shipment of whale meat due to be smuggled into the country.

The United States and other IWC members have tried to make illegal whaling and smuggling more difficult, but such illegalities show that, where profit can be made, science, conservation, and international law are of small concern to the whaling industry.

THE FUTURE

Recent years have witnessed some dramatic developments at the IWC. At the 1996 meeting, several countries announced their opposition to whaling under any circumstances. This was a controversial position to take—there were fears that it may prompt whaling nations to leave the IWC altogether—yet it reflected a moral view shared by millions of people worldwide.

Then, at the 1997 meeting, Ireland proposed limited commercial whaling in coastal waters, on the understanding that the meat could not be traded internationally and that all other whale hunts should stop. Unfortunately, the proposal was taken seriously and is to be discussed at the next meeting. Only time will tell how the IWC will evolve. But one thing is certain: the whales are not saved yet.

INDIGENOUS WHALING

In the past, some traditional human societies may not have survived without whale meat as a staple source of food. But whether or not they should still be hunting whales in the late twentieth century is a highly contentious issue.

The International Whaling Commission (IWC) rules currently allow indigenous or subsistence whaling in the Russian Federation, the United States, Greenland, and on the Caribbean island of Bequia, which forms part of St. Vincent and the Grenadines. Special quotas are agreed by the IWC, which allow certain communities to hunt a specified number of whales every year.

The numbers involved are quite small compared with the whales killed by commercial whalers and, ironically, some traditional communities have to live with the local legacy of commercially decimated whale populations. Nonetheless, indigenous whaling can still have a significant impact on whale populations and, perhaps more importantly, it has much wider implications for whaling in other parts of the world.

TRADITIONAL OR MODERN?

A strong argument against indigenous whaling, at least in some parts of the world, is that many "traditional" communities are actually quite modern. They no longer rely on whales for their survival.

A good example is a town called Barrow, on the northern

BEFORE 1926 *A Makah whaler holds two sealskin floats, filled with air. These were tied to the harpoon's rope to keep the harpoon afloat after being thrust.*

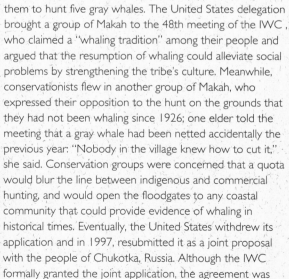

THE CASE OF THE MAKAH INDIANS

The Makah tribe lives in an isolated corner of America's Pacific Northwest, on Washington State's Olympic Peninsula. In 1996, its people made headline news around the world when they persuaded the United States government to apply for a quota that would allow them to hunt five gray whales. The United States delegation brought a group of Makah to the 48th meeting of the IWC, who claimed a "whaling tradition" among their people and argued that the resumption of whaling could alleviate social problems by strengthening the tribe's culture. Meanwhile, conservationists flew in another group of Makah, who expressed their opposition to the hunt on the grounds that they had not been whaling since 1926; one elder told the meeting that a gray whale had been netted accidentally the previous year: "Nobody in the village knew how to cut it," she said. Conservation groups were concerned that a quota would blur the line between indigenous and commercial hunting, and would open the floodgates to any coastal community that could provide evidence of whaling in historical times. Eventually, the United States withdrew its application and in 1997, resubmitted it as a joint proposal with the people of Chukotka, Russia. Although the IWC formally granted the joint application, the agreement was carefully worded to recognize the needs of the Chukotkan people, and at the same time making it virtually impossible for the Makah Indians to hunt gray whales for at least another year.

tip of Alaska. The local Inupiat and Yup'ik people have hunted bowhead whales in the region for more than 2,000 years and, even today, they are allowed an annual quota of more than 60 whales.

But the hunt is a sensitive issue. Not only is the bowhead an endangered species, but Barrow is no longer the primitive settlement it once was. Even the hunt has changed beyond all recognition, with no traditional hunting techniques being used. The whalers use motorized boats and rely on spotter planes to find the whales. They use heavy iron harpoons, including some that explode on impact, to do the killing. Then, since the modern local people do not rely on whale meat themselves, they generate a profit from their kills by selling carved baleen, or whalebone, to visiting tourists.

The Inuit are caught in the Catch-22 position of being swept up into an affluent,

ANNUAL QUOTA *A bowhead whale is landed by the local people of Barrow, in Alaska (below). An Inuit hunter (above left), using traditional methods, harpoons a narwhal in northwest Greenland.*

There is a passion for hunting … deeply implanted in the human breast.

Oliver Twist,
CHARLES DICKENS
(1812–70), English novelist

developed society while, at the same time, being expected to adopt the conservationist concerns of the outside world.

TRADITIONAL OR COMMERCIAL?

Perhaps inevitably, indigenous whaling has attracted the attentions of Japan and Norway. It is yet another possible loophole in the IWC rule book and, on many occasions, the two countries have attempted to expand

their own whaling operations under this guise. In particular, they repeatedly propose "small-type community-based coastal whaling" for towns and villages with a "whaling tradition." But in their efforts to deflect attention away from the commercial nature of the hunt, they ignore the fact that most of these "whaling communities" have not hunted whales for many years or, worse still, they began whaling as recently as the 1930s or 1940s.

Ultimately, the challenge is to balance the nutritional needs and genuine long-established cultural rights of native people with the importance of protecting whales. Of course, each case has to be judged on its own merits but, even at its best, the balance will always be a delicate one.

ABORIGINAL WHALING QUOTAS

Active Whalers	Species	1995	1996	1997
Russian Federation	Gray whale	140	140	140
West Greenland	Minke whale	155♦	155♦	155♦
	Fin whale	19	19	19
East Greenland	Minke whale	12	12	12
St. Vincent	Humpback whale	2	2	2
United States	Bowhead whale	68	67	66
Canada*	Bowhead whale	2	2	2

♦ Represents a total of 465 whales over three years (1995–97) with a maximum of 165 strikes each year.
* Not under IWC jurisdiction.

HUNTING SMALL CETACEANS

Tens of thousands—possibly even hundreds of thousands—of small whales, dolphins, and porpoises are hunted every year in seas and oceans around the world.

A motley collection of nets, knives, spears, rifles, and explosive harpoons are used by hunters. The animals are slaughtered for a variety of reasons: in most cases, to provide meat for human consumption or as scapegoats for badly managed fisheries, but they are also killed for crab bait and even to be ground into fertilizer or chicken food.

SUBSISTENCE HUNTING

Few communities genuinely rely on small cetaceans for their survival. But belugas and narwhals have been hunted in the Arctic regions of Canada, the United States, Greenland, and Russia for centuries. Their oil is used for lighting and cooking; their blubber and skin as a delicacy, known as *muktuk*; and their meat for human consumption and as food for sled dogs. The long tusks of male narwhals were originally used for tent poles, sled runners, and lance shafts (instead of wood), and were marketed as the horns of the legendary unicorn. More recently, until the introduction of trade bans, they have been sold as curios to tourists.

These days, about 1,000 narwhals are taken every year, mainly by the Inuit of the Canadian Arctic and West Greenland (many more are hit, but sink before they can be retrieved). The Alaskan Inuit take perhaps 200 belugas each year, apparently with no negative impact on the population. But more than 2,000 belugas are taken by the Inuit of the Canadian Arctic, West Greenland, and Arctic Russia, possibly with serious consequences for the future survival of the population.

COMMERCIAL HUNTING

In many parts of the world, small cetaceans are also killed commercially. Two of the largest commercial hunts take

---

THE FAEROESE PILOT WHALE HUNT

Long-finned pilot whales have been hunted around the Faeroe Islands, a group of Danish islands in the northeast Atlantic, for several hundred years. Entire pods of the 13–20 foot (4–6 m) whales are herded into sandy bays by men in small boats (below) during noisy drives that frequently take many hours. All the animals—including pregnant and lactating females with calves—are dragged ashore with steel gaffs, which are stabbed into their heads. They are killed with long knives and the meat is distributed free to local people (although some is also sold in supermarkets and occasionally appears on restaurant menus).

Faeroese hunting statistics date from 1584, with unbroken records from 1709 to now. During this most recent 287–year period, more than 250,000 long-finned pilot whales have been killed in about 1,700 different drives. The official number taken each year ranges from zero to a peak of 4,325, recorded in 1941. The average during the past decade has been about 1,200 whales a year. The Faeroese defend the hunt vigorously, arguing that it is a long-standing tradition and provides a free and welcome source of protein. But, in truth, it is no longer necessary in such a modern society with a relatively high standard of living.

HUNTED CETACEANS *Baird's beaked whales (far left) are one of 16 species still hunted vigorously off Japan for their meat. The narwhal (right), found in the Arctic, is prized for its spectacular tusk. Burmeister's porpoises (below right) were hard hit by Peru fisheries for their meat.*

place in Peru and Japan. Until quite recently, as many as 20,000 dolphins and porpoises were being killed every year by fishers in at least 60 ports up and down the coast of Peru. Dusky dolphins and Burmeister's porpoises were the hardest hit, but a number of other species also suffered severe losses. Sale of the meat for human consumption was a lucrative business and, as the fishers became increasingly greedy, more and more animals were being killed.

After a long campaign by national and international conservation groups, the Peruvian government finally banned dolphin hunting in 1990. But no effort was made to enforce the new law, and even the local police turned a blind eye. Eventually, in 1996, the government provided better enforcement and introduced long jail sentences and other strict penalties. But, although the scale of the problem has been reduced, the hunt continues.

In Japan, the situation is even worse. Japanese fishers actively hunt no fewer than 16 of the 21 species of small cetaceans in their coastal waters. Small coastal vessels, equipped with cannons that fire explosive harpoons, are used to kill about 50 Baird's beaked whales every year. Hand-held harpoons are thrown from small boats to catch Pacific white-sided dolphins and Dall's porpoises. And drive-fisheries (where fishers herd cetaceans into the shallows and kill them from shore) target striped, Risso's, bottlenose, and spotted dolphins; short-finned pilot and false killer whales; and a number of other species. In most cases, the meat is sold throughout Japan—often fraudulently as whale meat.

SCAPEGOATS

Fishers often blame whales, dolphins, and porpoises for damaging nets and for stealing or scaring away "their" fish. In retaliation, they kill them whenever they get the chance. There are even organized "culls"—sometimes with government approval or support—to wipe out as many small cetaceans as possible.

The largest hunt designed to reduce the level of competition with fisheries has been taking place in the Black Sea since 1870. During the worst period, from 1931–41, more than 50,000 dolphins and porpoises were being killed every year.

The hunting methods used are also cause for concern. In the 1920s and 1930s, the Quebec government literally bombed belugas from the air; in 1956, on the other side of the North Atlantic, the Icelandic government arranged for a United States naval vessel to use machine guns, rockets, and depth charges against killer whales; and in the mid-1980s, Alaskan fishers used explosives and firearms to wipe out local killer whales.

Despite this uncontrolled onslaught, there is often no direct evidence to support fishing industry claims of competition. Marine ecosystems are very complex, so it is actually very difficult to calculate the impact of small cetaceans (or any other predators) on fisheries.

Besides, the limited evidence that is available frequently discredits any claims of competition. In many places, for example, the fishers and small cetaceans are taking entirely different species of fish. The scarcity of fish in some other areas is more often the result of over-exploitation by the fishers themselves.

47

CAPTIVE CETACEANS

Is there any justification for taking cetaceans from the wild and keeping them in captivity—or is it done purely for financial gain?

Whales, dolphins, and porpoises are the undisputed stars of aquariums, marine parks, and zoos around the world. Every year, millions of people flock to see them "kissing" their trainers, fetching balls, jumping through hoops, performing somersaults, and making synchronized leaps in special choreographed shows.

Since the first bottlenose dolphins were taken into captivity more than a century ago, at least 25 species have suffered a similar fate. Today, Pacific white-sided dolphins, false killer whales, belugas, orcas, Irrawaddy dolphins, and many others are being held in a collection of netted pools, concrete tanks, and even hotel swimming pools. Although a few of these are bred in captivity, the vast majority are still being taken from the wild. During 1997, for example, a pod of orcas was driven ashore by fishers in Japan; half the struggling animals

were taken for sale to marine parks and aquariums.

It is not surprising that, for years, animal-welfare and conservation groups have been campaigning for a total ban on the capture of wild cetaceans. They would also like to see captive animals being released, where possible.

CRUELTY

The people who keep whales, dolphins, and porpoises argue that captivity is a simple trade-off. The animals lose their freedom and natural companions in return for escaping the two biggest problems of life in the wild: going hungry and being eaten. But animal-welfare groups argue that cetaceans

are completely unsuited to life in captivity, so keeping them in small tanks and pools is both immoral and cruel.

In the early days, a great many animals died during the first few hours or days after capture, and the remainder rarely survived more than a few months. Even today, when we understand more about their requirements, the transition from life in the wild to confinement can be most traumatic. Unknown numbers die in the process of transition.

Their final destinations include a bewildering range of establishments, from badly run zoos where they are poorly treated, to professional marine parks that provide the best care money can buy. The best establishments have coastal enclosures that are filled naturally with sea water and flushed with every new tide; they are often several acres in size and fairly deep. The animals are kept in carefully structured family groups, fed on varied and healthy diets, and have regular checkups by experienced veterinarians.

Unfortunately, such facilities are in the minority. Most consist of bare and

Fetters of gold are still fetters and even silken cords pinch.

English proverb

"AMBASSADORS?" *The argument is that facilities such as Vancouver Public Aquarium, British Columbia (above), and Sea World, Orlando, Florida (left and below left), provide the only opportunity most people will ever have to experience live cetaceans.*

featureless concrete tanks, with filthy water, and sometimes no natural sunlight. The animals are kept alone, or with unfamiliar company, and have to adapt to an unhealthy diet of dead fish. Sometimes, the captives simply cannot cope. They swim in circles, stop vocalizing, become aggressive or depressed, and may even harm themselves.

EDUCATION

One of the main arguments for keeping cetaceans in captivity is education. The protagonists claim that captive animals act as "ambassadors" for their wild relatives. But animal-welfare groups argue that few captive facilities make use of the great potential for educating and informing the public. If there is any educational element, it merely pays lip service to the concept of an "informative commentary."

Choreographed shows are particularly contentious. Trainers argue that they are educational, and that they keep the animals physically and mentally fit. But animal-welfare groups claim that they are just cheap entertainment for profit—and do little more than perpetuate the domineering and manipulative attitude we have toward nature. Moreover, the animals are often forced to perform— in some cases, they are even starved until they get it right.

At the end of the day, it is not essential to see whales and dolphins in real life to understand and appreciate them. After all, it is possible to learn about the Moon without actually standing on it. So while a live animal is more likely to trigger the emotions than words, photographs, or films, these could all be used as strong "second bests." Meanwhile, there are some exciting possibilities for the future, including computer technology that could simulate virtual-reality encounters with whales and dolphins where they really belong—in the oceans, wild and free.

CAN CAPTIVITY HELP TO PROTECT ENDANGERED SPECIES?

Despite many claims to the contrary, most marine parks and aquariums do nothing to protect endangered whales, dolphins, and porpoises. Simply keeping endangered species in captivity has no practical benefit and is not conservation. In fact, the opposite is usually true, because taking the animals from the wild can be a significant drain on local populations.

One particular species, however, is so rare that its only hope of survival may lie in semi-captivity. The Yangtze River dolphin, or baiji, is the world's rarest cetacean. Fewer than 100 are believed to survive in the wild. It faces so many threats—overfishing, dangerous fishing hooks, riverbank development, dam construction, pollution, and heavy boat traffic—that it is already doomed in its natural home in the Yangtze River, China.

So far, there has been little success in capturing a handful of the elusive survivors to establish a small breeding colony in the relative safety of a semi-natural reserve. But unless this last-ditch effort is successful, there are genuine fears that the baiji could disappear altogether within a decade. Even if this program is successful, whether or not the baiji (seen left) can ever be returned to the wild remains another matter.

THE IMPACT *of* FISHERIES *on* CETACEANS

The number of whales, dolphins, and porpoises drowned in fishing nets is staggeringly high—possibly amounting to millions of animals every year.

In the high-tech scramble to meet a growing world-wide demand for seafood, it is not only fish stocks that suffer—cetaceans and other wildlife are victims, too. Blame rests squarely on the introduction of increasingly destructive fishing methods and the sheer scale of many modern fisheries.

We are still a long way from identifying the precise impact of most fisheries on cetaceans, let alone enforcing practical solutions. But there is little doubt that urgent action is needed if we are to avoid disastrous population declines. In some cases, a simple modification to either the nets or the fishery management systems can have a positive effect. But far more drastic action, such as seasonal closures of some fisheries, may be the only effective long-term solution.

DRIFT NETTING

First used on a colossal scale in the mid-1960s, this is probably the most indiscriminate method of fishing ever

OVER-EXPLOITATION OF FISH STOCKS

Overfishing is one of the major conservation crises of the 1980s and 1990s and the sheer scale of many modern fisheries could be a major threat to whales, dolphins, and porpoises.

Put simply, fish are being taken faster than they can reproduce (see the harvest of dog salmon, below, in China). According to the Food and Agriculture Organization of the United Nations (FAO), 70 percent of commercially important marine fish stocks are fully fished, over-exploited, depleted, or slowly recovering. Yet the global fishing fleet—subsidized by many governments—continues to expand.

The impact of such intensive fishing on cetacean populations is largely unknown, but as the level of human competition for dwindling fish stocks intensifies, common sense suggests that there will be less available for cetaceans and other wildlife to eat. It certainly does not bode well for the future.

devised. A single drift net can be up to 30 miles (48 km) in length. Normally released at dusk, it is allowed to drift with the ocean currents and winds before being retrieved the following day, or several days later. It is virtually undetectable as it hangs in the water, and catches literally everything in its path. There are thousands of miles of these "walls of death" floating around the world's seas and oceans at any one time—more than enough to circle the Earth at the equator.

In recent years, the world

has slowly begun to wake up to the dangers of drift netting, and a UN resolution on nets longer than 1.6 miles (2.5 km) was finally agreed in December 1992. But even a net of this size poses a considerable threat to marine life and, of course, in defiance of the resolution, many fisheries continue to use much longer nets.

COASTAL GILL NETS

These are used in shallow, coastal waters. Made of the same nylon monofilament line as drift nets, and equally difficult for cetaceans to detect underwater, they either float at the surface or are anchored near the seabed. Cheap and durable, they are popular in many parts of the world. The

INDISCRIMINATE *An endangered vaquita is accidentally caught in a net, along with the target species. Cetaceans drown when they can't surface for air.*

FISHERIES *are keen to avoid trapping large cetaceans, such as the humpback caught in a fishing net (right), because they damage the nets. A recently trialled acoustical "pinger" that can be attached to the invisible nets holds great promise.*

gill-net problem stretches from New Zealand to Sri Lanka and from Canada to Britain, bringing death to a great many small cetaceans every year. It is even possible that dolphins and porpoises are attracted to the nets by all the trapped fish—and then become entangled themselves.

TUNA FISHING

The tuna-fishing industry has received a great deal of adverse publicity, and deservedly so, since it has killed more dolphins in the past 35 years than any other human activity. It is directly responsible for the deaths of some 6 to 12 million spotted and spinner dolphins, and several other species. Indeed, during the worst period, the 1960s and early 1970s, the industry was killing as many as half a million dolphins a year.

The problems began in 1959, when a new kind of net, the purse-seine net, was introduced to catch yellowfin tuna in the eastern tropical Pacific (a stretch of ocean extending from southern California to Chile and covering an area roughly the size of Canada).

It arrived at a time when yellowfin tuna fishers began to use the presence of dolphins to find their quarry (tuna and dolphins often swim together, but only the dolphins have to surface for air). "Fishing over porpoise," as the practice became known, increased tuna-fishing profits substantially but also sentenced

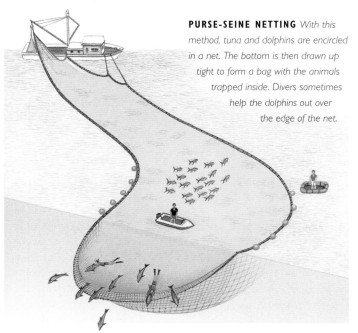

PURSE-SEINE NETTING *With this method, tuna and dolphins are encircled in a net. The bottom is then drawn up tight to form a bag with the animals trapped inside. Divers sometimes help the dolphins out over the edge of the net.*

millions of dolphins to slow, lingering deaths in the great canopies of loose netting.

After years of public outrage, the tuna-dolphin slaughter continues—but on a much smaller scale. The tuna-fishing industry is now governed by a variety of rules and regulations, ranging from special escape hatches in the nets (which allow trapped dolphins to be released) to the presence of official observers on tuna-fishing boats. The introduction of "dolphin safe" or "dolphin friendly" tuna, in 1990, was another step in the right direction, but there is still no independent enforcement scheme to check the truth of such claims. Already, less reputable tuna companies have been caught cheating.

In 1992, some governments signed the Agreement for the Conservation of Dolphins, known as the La Jolla Agreement. This sets dolphin mortality limits for each vessel participating in the fishery and has been an enormous success: the death toll has since fallen to about 4,000 dolphins per year. This is still too high and, worse still, there is mounting evidence to suggest that dolphins are being set on by tuna-fishing fleets in other parts of the world—so the problem could now be repeating itself.

MARINE POLLUTION

*A silent, insidious, and widespread killer, marine pollution seriously
threatens the future of cetacean populations around the world.*

Although seas and oceans cover more than two-thirds of the Earth's surface, they do not have an infinite capacity to absorb the plethora of waste products human activities create. Yet ever-increasing quantities of industrial waste, agricultural chemicals, untreated sewage, radioactive discharges, oil, modern plastic debris, and a huge variety of other pollutants are dumped directly into the sea—or slowly make their way there via rivers and atmospheric deposition. Once they are released into the environment, there is no way of recovering them and they can continue to cause harm for years or even decades.

The scale of the problem is hard to imagine. For example, it is estimated that no fewer than 63,000 chemicals are in common use worldwide—and as many as 1,000 new ones enter the market each year. These include DDT, aldrin, endrin, dieldrin, PCBs, lead, mercury, cadmium, zinc, and a motley collection of other organochlorines, heavy metals, and similarly dangerous things.

We are only just beginning to learn about the precise details of the damage such pollutants are able to cause. Measurements are difficult because the quantities involved are often unknown, their effects depend upon everything from the type of pollutant to the length of exposure, and most of the

DUMPING *Coal-mine waste (above) pours into the North Sea from Easington Colliery in Britain. The sea is discolored for miles around the outlet and beaches are black with coal waste.*

victims die at sea. But there is little doubt that their effect is often devastating and, already, some experts are predicting that pollution could be the most serious threat to the survival of cetacean populations in many parts of the world.

EFFECTS OF POLLUTANTS

Some pollutants are so toxic, or are present in such huge quantities, that they may cause immediate death. Others are more subtle in their effects, but nonetheless may be responsible for weeks, months, or even years of prolonged suffering.

In most cases, details are still unknown, but they are believed to weaken the animals, gradually causing hormonal imbalances, a lowering of disease resistance, brain damage and various neurological disorders, cancer,

liver troubles, and many other abnormalities and chronic health problems. They may even cause a lowering or total loss of fertility.

Minute quantities of toxins in the sea are picked up by marine plankton, which are then eaten by fish and squid, and these in turn are eaten by top predators, such as whales, dolphins, and porpoises. In this way, especially high concentrations of toxins build up in the bodies of animals at the top of the food chain, making them particularly vulnerable to the harmful effects of pollution. As the cetaceans get older, more toxins tend to accumulate in their blubber,

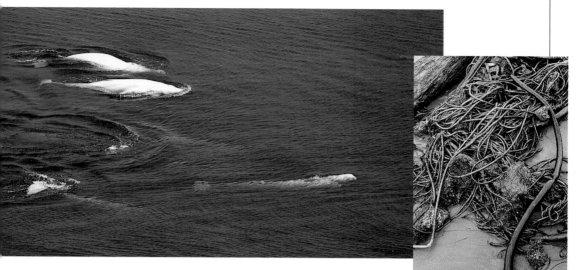

and in organs such as the liver. Much of this buildup is passed on from one generation to another: a lactating female, for example, delivers the toxins in highly concentrated doses to her young calf through her milk.

Species living in partly enclosed waters, such as the Baltic Sea, Black Sea, North Sea, and the Gulf of Mexico, or along polluted coastlines, are especially hard hit. Few places are entirely free of pollution, but in some areas, the animals are so highly contaminated that their bodies could be classified as toxic waste. Species living in the open sea, or in marine ecosystems far from human activity, fare little better: the tides, currents, and winds ensure that pollutants eventually reach even the

most remote corners of the globe. For example, bowhead whales in the high Arctic are believed to be exposed to heavy metals, chlorinated hydrocarbons, and a huge range of other pollutants.

ACTION IS NEEDED

Despite all the warnings, and high-profile campaigns by conservation groups, many governments choose to ignore the severity of the marine pollution problem. Some have improved their control of marine pollution—for example, strict regulations on organochlorines are in place in western Europe, North America, and some other parts of the world. But the majority continue to allow their waste to escape into the world's seas and oceans with barely a second thought, thinking only

HUMAN FOLLY *Belugas (above left) living in the Gulf of St. Lawrence have exceptionally high levels of mercury, PCBs, DDT, and a complex mixture of other pollutants in their bodies. The kelp (above) is badly contaminated with oil.*

in the short term, with a dump-it-now-worry-about-the-effects-later approach. The fact that many pollutants, even if they were banned today, would remain in the marine environment and still be dangerous for many years to come, makes such a cavalier attitude all the more risky.

One of the main problems is demonstrating a direct link between marine pollution and deaths, illness, or weakness in marine wildlife. It takes a long time before conclusive results can be obtained (if, indeed, they can be obtained at all) and yet, without them, it is difficult to convince governments that action is needed. But it is precisely because of all the unknowns that we should be especially judicious—and that means introducing pollution regulations and legislation before any more damage is done.

VISIBLE *While most victims die at sea, this dead hump-backed dolphin (left) is potent evidence of a toxic oil spill.*

HIDDEN DANGERS

As if hunting, pollution, and conflicts with fisheries

were not enough, whales, dolphins, and porpoises also have to contend

with a wide range of other threats.

Huge environmental problems, such as global warming and the depletion of the ozone layer, could ultimately have a detrimental impact on cetacean populations. Exactly how they might affect the world's waters, and their inhabitants, no one really knows. But there is intense debate about how global warming may be causing sea levels to rise, for example, and there are fears that ozone depletion may harm plankton.

Meanwhile, there are other, more imminent, problems. In particular, river dolphins face such a barrage of threats that it is hard to imagine them surviving beyond the twenty-first century. As well as hunting, pollution, and conflicts with fisheries—which are bad enough—heavy boat traffic, riverbank development, and dam construction all contrive to ensure that several river dolphin species may already be doomed in their natural homes. The Yangtze River dolphin, or baiji, is now so rare that the world's largest hydroelectric project at the famous Three Gorges, in Hubei Province, China, will probably be the final nail in its coffin.

HABITAT LOSS

Habitat loss is the single most important threat to terrestrial wildlife. Nowhere is entirely safe and, nowadays, there are few parts of the natural world that have not been altered, damaged, or destroyed. Worse still, it is not just a single species of animal and plant that is disappearing along with its habitat, but entire ecological communities.

The situation facing whales, dolphins, porpoises, and other marine wildlife is rather different, and perhaps more limited, but it is no less important. Land reclamation, commercial fish farming, coastal and riverbank development, disturbance from boat traffic, and the effects of land-based activities such as deforestation and river damming are all taking their toll.

Species living close to shore, or in rivers, tend to be the hardest hit, but human activities also reach far out to sea. The most frightening aspect of habitat loss is the

way it affects species that already have restricted distributions, or have such precise habitat requirements that they are tied to specific coastal or riverine areas and cannot easily "escape." The vaquita, river dolphins, Hector's dolphin, and Burmeister's porpoise, as well as several large whales that move inshore at certain times of the year to breed, are, obviously, particularly vulnerable.

But it is not just the direct effects of habitat loss. Degradation of the coastal zone can also damage the nurseries of fish and other wildlife that form the foundation of the sea's complex food webs. This could have a disproportionate impact on the productivity of the marine environment as a whole, and may yet prove to be one of the most serious long-term threats to cetaceans.

NOISE POLLUTION

The underwater world can be surprisingly noisy. Marine mammals, invertebrates, and fish make a medley of different sounds: the low-frequency moans of blue and fin whales, for

SHORE DEVELOPMENTS
such as a nuclear power station (left), affect cetacean habitats. The baiji's (above left) habitat faces a multitude of threats.

HABITAT CONSERVATION

The long-term solution is to protect entire habitats. No matter how many laws are passed to protect whales, dolphins, and porpoises, they will be useless if the animals have nowhere safe to live. Habitat protection means providing special sanctuaries, or marine reserves, in which they are guaranteed long-term protection from habitat degradation and disturbance as well as hunting, destructive fishing methods, and myriad other threats. Such reserves need to be sufficiently large to provide safe refuges in which the animals can feed and breed in safety. For example, the Irish Whale Sanctuary takes in all coastal waters around Ireland (above, Mizen Head Peninsula, Cork). This form of marine conservation is still in its infancy, lagging far behind conservation on land, but governments are slowly beginning to recognize the need.

example, are so loud that they can be heard for thousands of miles. A variety of natural events, from underwater volcanic eruptions to storms and heavy rain, add to the cacophony.

In recent years, human activities have added considerably to these natural sounds—and experts are seriously concerned about their impact on marine life. Coastal development, speedboats, jet skis, dredging, heavy shipping such as tankers and container ships, low-flying aircraft, military maneuvers, seismic testing for oil and gas, drilling rigs, sonar, and acoustic telemetry are all to blame for this increased noise pollution.

There has been relatively little research in this field, so the harm caused to whales, dolphins, and porpoises is largely unknown. Besides, it is difficult to evaluate because it is not just the loudness of a noise that is important, but also its frequency. Different species are believed to be more sensitive to some frequencies than others: in particular, larger cetaceans to low frequencies and small cetaceans to higher ones.

Even without a great deal of research, logic suggests that cetaceans are likely to be highly susceptible to noise pollution. After all, they live in a world of sound and rely heavily on effective hearing for communication, finding their way around, locating prey, and avoiding predators. Indeed, there is a growing body of coincidental evidence to suggest that, as well as directly interfering with all these day-to-day activities, extraneous noise may also have side effects such as reducing the animals' sensitivity to important sounds or causing them high stress levels.

NOISE POLLUTION
from oil drilling rigs may disturb the communication and navigation of cetaceans.

Meanwhile, experts fear that the increasing number of sperm whale strandings in Britain may be linked to seismic exploration in the deep waters west of the Shetland Islands. On the other side of the Atlantic, off the coast of Newfoundland, explosions from a nearby underwater drilling operation seem to make humpback whales more likely to blunder into fishing nets.

It has also been suggested that heavy shipping traffic, and its associated noise, may be one of the reasons the endangered northern right whale population is failing to proliferate (while southern right whales are increasing by about 5–7 percent annually).

Even if noise is not the prime reason for some of these problems, it is hard to imagine that it could be anything but another burden for whales to endure.

CARING *for* WHALES, DOLPHINS, *and* PORPOISES

Although whales, dolphins, and porpoises continue to be killed by myriad human activities around the world, there is still hope for the future.

It is only natural that, from time to time, everyone involved in whale conservation feels a sense of despair and helplessness. But progress is being made, albeit slowly, and the attitudes of governments and other key decision makers are gradually changing. In the past decade, there have been many success stories, from the establishment of the Southern Ocean Whale Sanctuary, surrounding Antarctica, to the passing of a new law to ban dolphin hunting in Peru. But there is still a huge amount to be done and there will always be setbacks.

WHAT'S THE ANSWER?

There are no easy solutions to most of the problems facing whales, dolphins, and porpoises. The issues are complex and there are often many vested interests involved. Solutions do exist, but they are often complex themselves and it may be many long years before they are put into effect.

The work undertaken by organizations such as the British-based Whale and Dolphin Conservation Society, the largest charity of its kind

dedicated to the conservation, welfare, and appreciation of cetaceans, is necessarily wide ranging. It includes anything from the development of good working relationships with key politicians to working toward a feeling of mutual respect and cooperation with local fishers.

It involves encouraging and assisting schoolchildren to take an interest in conservation, and focusing world attention on key issues, such as commercial whaling and destructive fishing methods. It entails producing action plans for saving endangered species, or populations, and developing realistic economic alternatives to hunting and killing. It involves undercover operations to gather important information on a wide range of illegal activities, improving the enforcement of existing laws and regulations, and much more. Above all,

constant vigilance is essential because, even when important progress has been made, it can always be weakened or revoked.

HOW CAN YOU HELP?

Concerned individuals really can make a difference. Without public support, there would be no money for conservation groups to carry out their vital work; and, without public pressure, there would be no incentive for key decision makers to take essential action. At the end of the day, there would be very few conservation success stories.

If you would like to help whales, dolphins, and porpoises, here are some ideas:
● Join a like-minded conservation group.
● Write letters of protest to key decision makers, organize petitions, and actively support conservation campaigns in other ways.
● Raise money through such activities as walks, bike rides, parachute jumps, and other sponsored events.
● If you have a special skill that might be of benefit to cetaceans (perhaps you are a journalist, a filmmaker, a printer, or a computer expert, for example), offer to donate some time and expertise.

EDUCATION *Whale lectures increase awareness and help secure the future of cetaceans.*

PROTESTS

Activists in Vienna form a bloody whale tail (above) to protest against Norwegian whaling. When Greenpeace monitored drift-net fishing in the North Atlantic, a French patrol tug (left) attacked Rainbow Warrior.

THE RAREST WHALES, DOLPHINS, AND PORPOISES IN THE WORLD

Species and Distribution	Population	Notes
Yangtze River dolphin or baiji		
Yangtze River, China (middle and lower reaches)	Fewer than 100 (possibly fewer than 50)	Now very little chance of rescuing this species. It is likely to become the first cetacean to become extinct in historical times
Vaquita or Gulf of California porpoise		
Extreme northern end of the Gulf of California (Sea of Cortez), Mexico	Fewer than 200	Has the most restricted distribution of any marine cetacean; most commonly seen around the Colorado River delta
Northern right whale		
Western North Atlantic (occasional records from eastern North Atlantic and eastern North Pacific)	Fewer than 320	Officially protected for more than 60 years, it has never recovered from being hunted almost to extinction by commercial whalers
Indus River dolphin or bhulan		
Indus River, Pakistan (mainly along 100 mile [160 km] stretch between Sukkur and Guddu barrages, or dams)	Fewer than 500	Since the 1930s, barrages have split the dwindling population into isolated pockets
Hector's dolphin *(inset picture)*		
Coastal waters of New Zealand (most common around the South Island)	Fewer than 4,000	The world's rarest marine dolphin, threatened mainly by incidental catches in coastal gill nets
Ganges River dolphin or susu		
Ganges, Meghna, Brahmaputra, and Karnaphuli river systems of India, Bangladesh, Bhutan, and Nepal	Fewer than 4,000	The population is split into two by the Farakka Barrage

Special note: population estimates are unavailable for many cetaceans, so it is possible that some poorly known species are even rarer than the ones on this list; at the same time, some species survive in slightly higher numbers, but are at least as endangered.

● If you own a company, or are employed by a company that you think may be interested in working with a conservation group for the benefit of cetaceans, investigate the possibilities for joint promotions and corporate sponsorship.

● Try to bring whales, dolphins, and porpoises into your life in as many ways as you can: if you are a teacher, tell your class about them; if you are a parent, tell your children; and if you are at work, tell your colleagues and anyone else who will listen. After all, the best way to gain support for cetaceans is through word of mouth.

CHAPTER THREE
ORIGINS *and*
ADAPTATIONS

They live in the midst of the sea, like fish;

yet they breathe like land species.

Histoire naturelle des cétacés,
BERNARD GERMAIN LACÉPÈDE (1756–1825), French naturalist and author

EVOLUTION *and* RADIATION

*In the 50 million years of their history, whales
and dolphins have shown many adaptations.*

ANCIENT FORMS *The fossilized skull
and jaw (above) of an extinct dolphin.
A triangular tooth (left) from an archaic
whale has cutting points like a shark's.*

Mammals are the dominant large land animals of the modern world. But, for about two-thirds of their evolutionary history, while dinosaurs dominated the Earth, they were small and probably rather insignificant. The earliest mammals, from about 210 million years ago, were shrew-sized species now known mainly from fossils of tiny teeth. There is little fossil evidence of large mammals until about 65 million years ago. Perhaps dinosaurs were so successful that there was no place in the world's ecology for larger mammals with the lifestyles familiar today.

After the mass extinction of dinosaurs 65 million years ago, mammals diversified widely and rapidly. They became the dominant land animals, with an average size perhaps that of a cat. Size, however, can vary from tiny (shrew) to enormous (larger than an elephant) for some extinct land mammals. Mammals diversified in feeding habits, including many specializations on the basic pattern of herbivore or carnivore. New environments were invaded worldwide, from seashore to mountains. Some mammals moved underground, and others to the treetops or to the air. Several groups took to water.

ORIGINS OF WHALES

The ancestors of whales and dolphins are known from 50-million-year-old fossils found in India and Pakistan. These fossils represent small, perhaps dolphin-sized, amphibious mammals that lived in a shallow subtropical seaway—the Tethys Sea, between the Indian subcontinent and Asia. Later, continental drift pushed India into Asia, closed the Tethys Sea, and upthrust the Himalayas. Ancient marine rocks containing whale bones are now exposed on land.

Among the archaic whales, called archaeocetes, is *Pakicetus*, found in sediments deposited in river or shallow marine settings. Skull bones, teeth, and a lower jaw point to a small cetacean, perhaps less than 6½ feet (2 m) long, which had a limited ability to hear underwater. Other early archaeocetes retain a pelvis and hind limbs, but probably swam by up-and-down movements of a strong tail, as do living cetaceans. Overall, these animals mark an early transitional amphibious stage in the shift from land to sea.

The land-dwelling relatives of early cetaceans were even-toed hoofed mammals, the cud-chewing artiodactyls. Supporting evidence for this comes from studies on living species, particularly on DNA, chromosomes, blood composition, and soft-tissue anatomy. So, sheep and cattle are among the closest living relatives of whales and dolphins.

ANCESTORS *The forebears of
cetaceans lived on land and
may have been dog-sized
animals that looked like
Mesonyx (right).*

for position" in the world's
waters, so that some groups of
cetaceans diversified as others
declined to extinction. About
12 to 15 million years ago, for
example, the first true dolphins,
porpoises, and white whales
appeared. After that, some
groups that were more
primitive declined and gradually disappeared. These
changes may also relate to
changing food resources
and marine circulation.

The last main
turnover among
cetaceans was about four
million years ago, about
the time our ancestors first
walked upright in Africa. As
climate gradually deteriorated
before the ice ages, the last
groups of archaic mysticetes
and odontocetes disappeared,
leaving a fairly modern
cetacean fauna.

To identify the exact
ancestors, we must turn to the
fossil record. One extinct
group, the mesonychids,
appears to include the ancestors of archaic whales. These
comprise a range of mostly
medium-sized (perhaps dog-
sized) animals that lived in
ancient Asia, Europe, and
North America. They showed
diverse feeding habits. Perhaps
cetaceans evolved when one
line of mesonychids took to
feeding on fish in rivers or
estuaries, rather like otters.

RADIATION

During their early history,
about 50 to 35 million years
ago, archaeocetes diversified
and spread from the tropics to
temperate waters. Skull form
became more complex, to aid
underwater feeding and hearing, and the hind limbs were
reduced to small vestiges.

About 35 to 30 million
years ago, the ancestors of
baleen whales and of toothed
whales and dolphins evolved
rapidly. These were the first

mysticetes and odontocetes.
Mysticetes showed a new
feeding style, that of filtering
their prey from the water.
This behavior allowed whales
to crop the extensive food
resources of the polar oceans.
Early fossil mysticetes filter fed
using sieve-like teeth. Baleen
evolved by about 30 million
years ago, and has been a key
feature of mysticetes since.

Fossil odontocetes from
about 30 million years ago
probably echolocated, using
high-frequency sound to navigate and detect prey. This
behavior has been a feature of
odontocetes since then.

Once the basic "modern"
behaviors of filter feeding and
echolocation appeared, cetaceans diversified dramatically
from about 25 million years
ago toward modern days.
They ranged from poles to
tropics, from near shore to
open ocean, from surface to
great depths, and even invaded
fresh waters. But, there was an
ongoing ecological "jostling

BODY SHAPE

Major changes in body shape
probably occurred 50 to
35 million years ago during
the time four-legged amphibious proto-whales evolved into
fully oceanic forms. Hind legs
became smaller and were
eventually lost, and tail flukes
developed. Though there is no
sure fossil evidence, we can
guess that, at this time, external ears and body hair disappeared and perhaps the dorsal
fin evolved. The forelimb
changed into a flipper with an
inflexible elbow. Later, early
odontocetes probably developed the bulbous forehead
and single blowhole seen in
living species. Odontocetes
also evolved a wide range of
upper-jaw forms, from short
and wide to long and narrow.
Among baleen whales, throat
grooves probably appeared in
rorqual lineages within the
past 10 million years.

MAMMALS *or* FISH?

Cetaceans are fast-moving, fully aquatic carnivores. They have a streamlined and maneuverable body propelled by a tail rather than by limbs.

Cetaceans show many fish-like features that are adaptations to life in water. These are over-printed on basic, underlying mammalian structures.

MAMMAL VARIATIONS

Cetaceans have a body form that lacks the projecting parts seen in most mammals. The skin is smooth, with hair present only as facial bristles in young cetaceans and a few adult baleen whales. The expressive "face" typical of land mammals is largely obscured by blubber. Most mammals have forward-facing nostrils, but the equivalent blowhole in living cetaceans is on top of the head. Farther toward the back, dimples on the sides of the head mark the position of the ears, but there are no external ear pinnae (flaps or lobes). Most of the cetaceans have a short and rather inflexible neck, and head movement is limited.

Behind the head, forelimbs are fin-like, without visible elbows and fingers. Generally, there are only small bony limb remnants, hidden far inside the body, with no hint of legs. The genitals don't protrude, other than for sexually active males, and there is no scrotum. Adaptation to aquatic life is a trade-off; some mammalian features cannot be lost fully but must be retained in modified form.

Perhaps the most obvious fish-like features are the dorsal

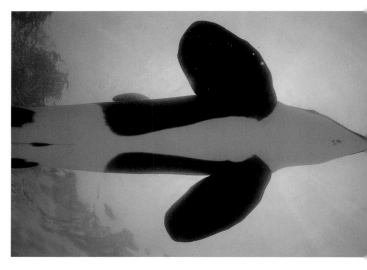

FULLY ADAPTED *The fin-like fore-limbs (above) of an orca have become pectoral fins, with no elbows or fingers.*

fin and tail flukes. These are both new structures created by evolution. Presumably, the dorsal fin is a hydrodynamic feature used in swimming. But in some cetaceans, the dorsal fin is small, and in a few it is absent. It is not clear how this sort of variation affects swimming ability. Horizontal flukes are present at the tip of the tail. These flukes are supported only in the midline by the hindmost tail vertebrae, and otherwise are constructed of tough non-bony tissues.

CETACEANS AS MAMMALS

Most mammals are active land-dwellers with a high metabolism, and a constant warm body temperature. Body structures match these basic adaptations. Breathing air is fundamental; the nose is well developed, and lung breathing is helped by a diaphragm and thoracic ribs. A four-chambered heart pumps blood to the lungs and

body. The genitals and anus are separate. Mammals carry the fetus in the uterus, suckle the young, and show complex patterns of reproductive and parental care. Cetaceans show all these features. More subtle mammalian features are also important. For example, the ear contains three tiny bones, the lower jaw is composed of a single bone, and the verte-brae in the backbone are separated into different regions from front to back.

Cetaceans inherited their basic swimming movements from distant land-dwelling ancestors. Fast-moving land mammals flex the body up and down, in contrast to the side-to-side movements of many reptiles and amphibians. Such movement is linked with rapid fore-and-aft motion of the limbs, seen well in a

galloping or bounding mammal. Cetaceans also flex the spine vertically, although they don't use fore- and hind limbs for movement. Rather, powerful muscles above and below the vertebrae give rise to tendons that run back, via a narrow peduncle, to power the tail flukes. Overall, propulsive muscles, tendons, and vertebrae act to beat the flukes up and down, providing thrust. Most cetaceans can also bend the body sideways.

Different cetacean species have different sizes and distributions of swimming muscles, and different shapes of flukes. Also, there is a wide range of swimming speed and maneuverability among species. Dall's porpoise, for example, is a remarkably fast oceanic odontocete, while some of the river dolphins, such as the boto, are slow yet very maneuverable. Broad correlations between body structure and swimming performance are not well understood.

OTHER MARINE MAMMALS

Demands of aquatic life have been solved in different ways by the other two marine

MAMMALIAN *features of a harbor seal (below) include fur and facial whiskers.*

VESTIGES *Hairs on the upper jaw of a gray whale calf (right) are its only remaining body hair. The dimple on the bottlenose dolphin's head (below right) is a vestigial ear opening.*

mammal groups: the pinnipeds (walrus, sea lions, and seals) and the sirenians (manatees and dugong).

All pinnipeds are powerful swimmers, and are superbly adapted to life in the sea, yet they have not made a total break from land. All their feeding is done at sea but they have to come ashore (or, in some cases, onto ice) to molt, rest, and breed. It is this need to return to land that accounts for the more typically mammalian features found in pinnipeds, but lost in cetaceans or sirenians. In particular, pinnipeds are unusual in having four webbed limbs, armed with claws at the ends of all the digits, and (in most cases) a dense coat of fur. The fur seals and sea lions can even use all their limbs to walk around on land (in true seals, the hind limbs are greatly modified for swimming, and cannot support the body).

Unlike cetaceans, seals have a social system which is clearly adapted for the land: smell, body posture, size (sexual dimorphism), and development of hair (such as the "mane" in some male fur seals and sea lions) all play a behavioral role.

Sirenians are fully aquatic. They also lack external hind limbs and have tail flukes. But sirenians are quite different from cetaceans in body form and habits. These slow-moving shallow-water herbivores feed on sea grass, and live only in warm waters.

What is more enthralling to the human mind than this splendid, boundless, colored mutability!—life in the making?

Adventures in Contentment,
DAVID GRAYSON, (1870–1946),
American journalist

LIVING UNDERWATER

Whales and dolphins show dramatic adaptations for life underwater. Changes in almost every body system make them quite unlike other mammals.

Land-dwelling mammals have load-bearing limbs to support and move the body. Because cetaceans live in water, a buoyant medium, limbs are not needed for support or propulsion.

Thermal and vascular adaptations of cetaceans are quite distinctive. Like other mammals, cetaceans maintain a constant warm internal temperature. Blubber insulates against the chilling effects of water and heat-conserving countercurrent exchange systems reduce heat loss in blood that circulates near the surface. Because of their body size and proportions, cetaceans have a problem getting rid of excess heat generated by exercise. Many species use countercurrent exchange systems in the flippers, dorsal fin, and flukes to radiate or conserve excess heat.

Land mammals need water to maintain the correct salt balance in the body. So, how do cetaceans deal with water and salt? Ingestion of salt during feeding, and from swallowed water, seems unavoidable, but we don't understand salt-balance mechanisms well. Probably, the complex cetacean kidney helps in the excretion of excess salt.

SLEEPING WITHOUT DROWNING

Sleep for humans means reduced muscle activity and reduced consciousness. Blood pressure and breathing rate drop, and the eyes close. Cetaceans apparently don't sleep quite the same way. In aquariums, animals rest quietly near the surface, or at the surface with the blowhole exposed. The breathing rate is lower than usual, and the eyes may close. But, breathing still seems to be voluntary, pointing to a wakeful level of consciousness. Anecdotal evidence points to some species, particularly the sperm whale, as sleeping soundly for long times at the surface, so much so that they can be a hazard to shipping. Some species rest during the day, but are more active at night.

EVOLUTION OF THE BLOWHOLE

We can breathe through either nose or mouth, but cetaceans breathe only through the blowhole (nostril). Cetaceans have effective structures to close the blowhole underwater. In mysticetes, the two blowholes are blocked by large nasal plugs. During breathing, the plugs are retracted by fast-acting muscles that originate on the upper jaw in front of the blowholes. (In humans, the same muscles lie between the lips and nose.) The action is remarkably fast, occurring in the brief interval that the whale breaks the water.

Odontocetes have a more complex blowhole, because the nasal passages are modified for sound generation (see p. 76). Two nasal plugs are still present, but they are buried far inside the head where they close off the more internal parts of the nasal passages. The single, external blowhole is a neomorph (or new

SLEEP *A beluga with blowhole open (top). A southern right whale (left) rests quietly at the surface of the water.*

THE BENDS AND NITROGEN NARCOSIS

Deep diving is hazardous for air breathers. Narcosis, a narcotic condition of stupor or insensibility, is caused by breathing air under pressure. And the bends (decompression sickness) is caused by surfacing too quickly from the depths, so that dissolved nitrogen suddenly bubbles out of the blood causing pain in joints and tissues. Yet although some cetaceans, such as the sperm whale (above left) and ziphiids, dive deep and long, with rather fast descent and ascent, they don't appear to be affected. Mechanisms that prevent the problems may include: diving with rather small volumes of air, thorax and lung collapse at depth (air squeezes from lungs into the windpipe, thus reducing nitrogen absorption), rapid transmission of nitrogen from blood to lungs at the end of a dive, and reduced circulation of blood to the muscles.

structure), which is an evolutionary change unique to odontocetes. The odontocete airway, unlike that in humans, is quite independent from the mouth and throat. Part of the larynx fits into the nasal passages at the back of the palate, so that food processing and breathing are separate.

MAINTAINING BUOYANCY

In most mammals, the body's center of buoyancy lies forward of the center of gravity, so that the animal floats with the head raised. Marine mammals, however, have a lifestyle that requires the body to be submerged except when breathing. Head and body orientation can be changed by ballasting (weight redistribution) or by hydrodynamics (swimming movements). Ballasting can involve lightening some bones, as seen in the porous and fat-filled vertebrae of some mysticetes. Alternatively, additional dense bone can be deposited, for example, in the ribs.

Ballasting is understood poorly for living cetaceans, but has been studied in some fossil species and also in sirenians (manatees and dugong). It seems to be linked with slow swimming speeds. For faster marine mammals, swimming movements maintain appropriate head and body posture. It has been argued that the spermaceti in the forehead of the sperm whale is used to help maintain buoyancy, but it seems more likely that the forehead has a function in sound transmission.

SWIMMING MODES IN CETACEANS AND FISH

Fast-moving fish and cetaceans are superficially similar: in both, the body is streamlined to reduce drag, fins are used to maneuver and stabilize, and a crescent-shaped tail beats to provide thrust.

In detail, there are many differences between cetaceans and fish. Fish show many more

INSULATION *A layer of blubber and muscle such as this one from a fin whale (left) insulates cetaceans against the cold.*

swimming techniques. Varied body undulations (including eel-like wriggling) may be involved. Cetaceans do not use rhythmic movements of the limbs as their main method of propulsion, whereas many fish and swimming land mammals (such as humans) do. Instead, cetaceans move the tail up and down, rather like body movements of a running land mammal (see p. 63).

FEEDING—BALEEN WHALES

Filter feeding is linked to many aspects of baleen whale structure and behavior, such as the huge head and mouth, large body, migration, and polar feeding.

Baleen whales are unusual mammals. Although enormous, they feed on tiny prey, filtering food from huge volumes of water in a remarkable harvesting operation.

BALEEN PLATES

Mysticetes (baleen whales) filter feed using baleen plates that hang from the roof of the mouth. Each baleen plate is thin, wide, and long, shaped like a narrow-based triangle, and formed of tough, flexible, organic material similar to hair or fingernail. Each plate grows down from the gum.

A plate is formed internally of tiny, elongate tubules that are exposed as hairs where the plate is worn along its inner edge. Different species have varying numbers, sizes, and colors of baleen plates (light, dark, or mixed).

Although it originates from the gum, baleen is not related to teeth. Teeth do occur in embryonic mysticetes, but they are resorbed before birth, leaving mysticetes with no teeth in either the upper or lower jaw. Fossils show that baleen whales arose from toothed ancestors. The geologically oldest mysticetes had well-developed teeth, perhaps associated with baleen; but by approximately 30 million years ago, no trace of teeth remained in many of the mysticete groups.

GULPING

Baleen whales show two main feeding strategies, "gulping" (or "swallowing") and

HUMPBACK FEEDING

Humpback whales show remarkably diverse feeding patterns. Feeding may involve lunges through prey shoals by a single whale, or synchronized group feeding, in which several humpbacks strike prey shoals together. They may even "herd" scattered prey into clusters that can then be caught more easily. A whale may swim on its side for better maneuverability, or circle around, disturbing the water and causing frightened prey to cluster. The whale then lunges through the shoal of food. In bubblenet feeding (left), humpbacks produce columns of bubbles that also cause prey to cluster tightly. A single, large, bubble net, produced by a continuous spiral of exhaled air, is believed to work just like a conventional fishing net. One or many whales may circle underwater, and then swim through the center of the net to catch their clustered prey.

"skimming." Gulping is common among rorquals. These whales have many external grooves, or pleats, below the mouth and throat. A gulping whale lunges at shoals of prey with its lower jaw dropped and the mouth wide open, and engulfs a huge volume of water and prey. Throat, or ventral, grooves allow the mouth cavity to increase greatly, forming an expanded pouch below the throat and thorax. With the mouth closed, water is forced out across matted baleen hairs and between baleen plates, so that food is trapped. The food is then swallowed.

Other features are linked with gulping. The lower jaws are bowed out, to increase mouth volume. Jaw-closing muscles have unusual orientations on the skull, to help close the widely opened jaw. Upper jaw bones are only loosely attached to one another, perhaps to reduce stress across the skull. Some specialized behaviors, such as bubble-net feeding (see box), are also associated with gulp feeding.

GIANT MOUTHFULS
As the humpback whale (above) lunges to engulf a shoal of fish, throat pleats expand and then contract to force water through the baleen filter hanging from the upper jaw. The arched upper jaw and baleen of this southern right whale (left) are clearly visible as it skims for food.

SKIMMING

Right whales feed differently, moving slowly at or near the surface, and forcing a stream of water across their very long baleen plates to skim off food. Feeding is more or less continuous, in contrast to the short, dramatic bursts of gulpers. Unlike balaenopterids, the narrow upper jaw is remarkably arched to hold the long baleen, giving the skull a most unusual appearance. There are no throat grooves.

Some balaenopterids use both skimming and gulping techniques, although most are gulpers. The gray whale is mainly a bottom feeder. It stirs up mud or silt, then filters small bottom-dwelling invertebrates from the slurry. It may also suck up prey.

Individually and overall, mysticetes eat a wide range of food. In the Southern Ocean, the main prey is the small, shrimp-like crustaceans called krill (euphausiids). Krill occur mostly at shallow depths, where they feed on small algae. It is not clear how baleen whales find krill swarms, but smell is possibly used. Copepods (another group of small, free-swimming crustaceans) are also an important source of food.

Northern mysticetes take a broader range of food, including krill, copepods, and amphipods (the group that includes sand hoppers). Schooling fish, such as herring and capelin, and mollusks, including squid and pteropods, also are taken.

Antarctic minke whales with body weights around 6.5 tons (6 tonnes) eat an average of 80–260 pounds (40–120 kg) of krill per day during a 120-day feeding season. Consumption for the larger sei whales, which mature at about 20 tons (18 tonnes), is less certain. Estimates are as high as 1,600 pounds (720 kg) of food per day, but they are known to eat at least 220–440 pounds (100–200 kg) daily. During a 130-day feeding season, bowhead whales, which may exceed 55 tons (50 tonnes), generally consume an estimated 1,100–3,300 pounds (500–1,500 kg) of food daily.

67

FEEDING—TOOTHED WHALES *and* DOLPHINS

Toothed whales and dolphins characteristically use echolocation as a feeding tool.

TOOTH VARIATIONS *A few examples of the teeth of odontocetes: (from left) the Ganges River dolphin, the harbor porpoise, the spiral tusk of a male narwhal, the sperm whale, and the strap-toothed whale.*

Toothed whales and dolphins (odontocetes) hunt and eat relatively large, single prey. This is a more generalized and more adaptable feeding method than the filter feeding of baleen whales. Feeding adaptability explains the greater diversity of odontocete species and range of habitats. The proportions of jaws and skull in different odontocetes hint at their feeding habits. Generally, long, narrow jaws allow a fast-closing forceps-like action, but not such a strong grip. Conversely, short and usually broad jaws provide power rather than fast action.

Ancient fossil odontocetes had triangular teeth with multiple cutting points. Over time, these teeth evolved into the simplified conical teeth of most odontocetes seen today. Occasionally, delicate "tusks" projected from the tip of the jaws. Some fossils suggest rather different feeding habits from those of living species. For example, an extinct bizarre dolphin from Peru may have been a suction feeder like the modern walrus.

Most odontocetes now have uniformly simple conical teeth that vary little from front to back. These homodont teeth probably function just to grasp food, in contrast with the more varied teeth used for complex food processing in most mammals. Yet there are interesting anomalies within this general pattern. The teeth of the rough-toothed dolphin, for example, have a wrinkled surface. It is not clear how such rough surfaces function. In the harbor porpoise, the teeth are spade-shaped and compressed from side to side; presumably, they shear food rather than merely holding it. Orca teeth are large, robust, and rooted in strong, short jaws adapted to bite chunks of meat out of large prey.

Odontocetes commonly have more teeth than usual for mammals. Sometimes, there are scores of teeth in each side of the jaw, as in the common dolphin and the franciscana. Squid-eating species, on the other hand, often have greatly reduced numbers of teeth.

VARIETY OF FOODS

Varied jaws and teeth among odontocetes suggest that different species have evolved

FAST GRABBER *The Amazon boto's long, powerful jaws (left) hold numerous rough-surfaced teeth. At the back of the jaw, these are quite wide and low.*

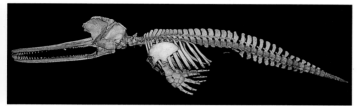

to specialize on different sorts of food, thus reducing competition. Strangely, many odontocetes are quite flexible in their choice of food. For example, spinner dolphins in the open ocean feed on mid-level fish, squid, and shrimp, but in shallow seas they may eat bottom-dwelling and reef organisms. Striped dolphins take mainly squid in some parts of their range, but mid-water fish in other regions.

A few odontocetes clearly are specialists with no close competitors. One, the Ganges River dolphin, has very narrow, long jaws with interlocking needle-like teeth. These jaws allow fast grabbing of mobile prey. The Ganges River dolphin usually swims on its side, perhaps for better maneuverability.

Male beaked whales (family Ziphiidae) commonly have one pair of highly distinctive protruding lower teeth, but females lack large, functional teeth. In males, teeth originate at the jaw tip (Cuvier's beaked whale), or well back along the jaw (Sowerby's whale). Male ziphiids often carry elongate subparallel scars on the body, apparently from fighting.

In the strap-toothed whale, a broad tooth from each side of the lower jaw curves up and over the upper jaw, greatly limiting the extent to which the mouth can open.

Given other sophisticated behavior in cetaceans, it is no surprise that cooperative hunting occurs in odontocetes. Groups of tucuxi (found in South American rivers and coastlines) may encircle fish schools in a coordinated attack. Small groups of orcas have been known to harass and attack individual mysticetes, working cooperatively to remove the soft tissues around the mouth. And in estuaries along the eastern shore of North America, bottlenose dolphins engage in a form of coopera-tive hunting (see p. 93).

SPERM WHALES FEEDING

The sperm whale is the largest odontocete and, with its strange teeth and huge face and forehead, is one of the more peculiar. Peg-like erupted teeth in the narrow lower jaw fit into sockets in the broad upper jaw. Upper teeth are vestigial and buried in gum tissue, but fossils indicate that in ancestral sperm whales, the upper teeth were functional.

Sperm whales routinely dive (right) to feed in very deep water, where they can take giant squid. These rapid-growing squid may have mantles more than 6½ feet (2 m) in length, and tentacles more than 32 feet (10 m) long. Sperm whales are believed to use echolocation to find their prey and probably use suction feeding to take the squid. (Whales with deformed jaws have still been able to eat, which indicates that teeth don't play a critical role.) There is no conclusive evidence of fierce deep-sea fights between squid and whales. Indeed, whales have such a size advantage that the contest is very one-sided. Less dramatically, medium-sized squid and, for near-shore sperm whales, a range of fish are also important as food.

AVOIDING PREDATORS

Over the millions of years of their existence, cetaceans have had to evolve strategies to deal with the attacks of their predators.

The natural predators of cetaceans are orcas and sharks, while in the Arctic, polar bears (above) may prey on stranded belugas. Close relatives of orcas, such as false killer whales and pygmy killer whales, have also been observed attacking other cetaceans. No cetacean species (with the possible exception of orcas themselves) is safe from attack: even blues or rights, the largest of the whales, may be attacked and killed.

FORMING GROUPS

The most basic defensive strategy cetaceans have evolved is to form a group. The detection of predators is enhanced by the combined senses of sight, hearing, and echolocation of all the group's members. In a large group, an individual's chances of being preyed on are considerably lower than if it were solitary, and group members can help to drive off attackers. Groups of oceanic dolphins, such as spinner and spotted dolphins, sometimes number thousands.

Why don't all species form large groups for defense? If their food supplies were large and predictable enough, they would be able to do so. But the prey of many cetaceans occurs only in small patches, and in large groups, group members might become rivals. So the group size of most species depends greatly on the ecology of their prey. Coastal bottlenose dolphins,

for example, feed on small schools of dispersed prey over a restricted range. They often form groups of fewer than 20 animals, which may offer optimal benefits in terms of both defense and foraging. In deep water away from the coast, where their food occurs in larger schools, but where there are also many sharks, the same species forms groups of several hundred individuals.

There is no doubt that in many parts of the world sharks prey heavily on dolphins. Near Sarasota, Florida, 20 percent of bottlenose dolphins bear shark wounds, while in Moreton Bay, Australia, more than a third of bottlenose dolphins show evidence of shark attack. There are occasional reports of groups of dolphins harassing or attacking sharks,

but the often-heard opinion that dolphins lord it over sharks is really a myth. Most scars seen on dolphins and other toothed whales, however, are more likely to result from aggressive interactions between the dolphins or whales themselves.

FIGHT OR FLIGHT?

Different strategies may be used when whales are under attack from predators. They may flee, or if they are slow-swimming or have young, they might stand and fight. Slow-moving sperm whales sometimes adopt a formation named after a type of flower, the "marguerite." Adult whales form a circle, with their heads toward the center and their flukes (their defensive weapons) facing outward. Young or wounded

FLOWER POWER *In the so-called "marguerite" formation, these sperm whales have arranged their bodies in a protective circle around calves or wounded adults as a pod of orcas attacks. Few predators would brave the thrashing tails or flukes turned against them in this way.*

SAFETY MEASURES *A female humpback (left) will shield her calf with her own body. Spinner dolphins (below) in Lanai, Hawaii, seek safety in numbers.*

animals are placed within the circle. It has been reported that on one occasion, whalers approached a marguerite formation around a wounded sperm whale, and harpooned almost all the members of the group, which refused to abandon their companions. Deep diving is sometimes used by sperm and other whales to escape attackers.

Many species of toothed whale are known for their "care-giving" behavior, in which individuals stand by and even defend a sick or threatened companion. Such altruism usually has a solid biological basis in genetic relationships, with such companions likely to be either closely related individuals, or females who could be bearing related offspring.

Baleen whales form smaller groups than do toothed whales, and have less cohesive and lasting social bonds. With the exception of mothers defending calves, there is no firm evidence that baleen whales will defend others from attack. Many humpbacks and other whales show fluke scarring, presumably from unsuccessful orca attacks.

Humpbacks have been seen to react to orca attacks with violent fluke thrashing,

turning, and rolling. Unless it responds very vigorously, there may be little that a solitary baleen whale can do to defend itself against such a formidable, coordinated attack. Yet not all baleen whales react in this way. A Bryde's whale killed by orcas in the Gulf of California seemed only to try to out-swim them. A gray whale with calf, attacked by orcas, also sought only to escape. Perhaps fluke thrashing is as likely to injure the calf as its attackers. Yet a southern right whale cow

Self preservation is the first principle of our nature.

A Full Vindication,
ALEXANDER HAMILTON
(1757–1804), American statesman

whose calf was badly mauled by a great white shark at a South Australian breeding ground, herself had mauled flukes—possibly from defending her calf.

OTHER TACTICS

An extraordinary defensive behavior seen in sperm whales shows a curious parallel to their favored prey—squid—which escape through clouds of ink. When threatened, sperm and pygmy sperm whales sometimes emit blinding clouds of feces. One wonders if the tactic comes from millions of years of hunting squid. A more passive form of camouflage is the countershading seen on many smaller cetaceans, which makes them more difficult for sharks to see.

Other defensive behaviors include making mock charges; puffing themselves up; blowing loudly; making threatening gestures, such as lobtailing and breaching; or a mother subtly placing her body between her calf and an approaching vessel.

FLURRY OF WATER *If threatened by a boat, an orca may try to intimidate the aggressor by lobtailing (left).*

SENSING *the* UNDERWATER ENVIRONMENT

*Water is a dense medium to which the sensory systems
of cetaceans have adapted remarkably well.*

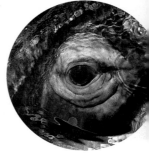

Cetacean sensory systems are well adapted to detect water-borne signals, which have unexpected characteristics. For example, sound travels at about 1,120 feet (340 m) per second in air, but in water it is about 4.5 times faster. Yet, cetaceans can detect and interpret this sound. Dissolved scent moves fast in air, but slowly in water, and a mammalian nose is not structured for sniffing water.

Light is important in surface waters, but even here, eyes must work with blue rather than a full range of color. Water provides buoyancy, which partly counters gravity, and allows much more three-dimensional movement. This, in turn, must be monitored in ways land mammals rarely have to do.

IMPORTANCE TO WHALES

Sound (see pp. 74 and 76), sight, and touch seem to be as important to cetaceans as to other mammals. These senses function in navigation, communication, feeding, and breeding, and work with all the other senses to monitor the environment continually. But other senses are understood in only a general way, because of the lack of detailed study. We know or guess that relevant detectors are present and are used, but we don't understand exactly how they work. These other senses include taste, smell, and the detection of gravity, acceleration, pressure, temperature, and magnetic fields.

SIGHT

The cetacean eye functions in a setting where temperature and pressure can change rapidly, where light levels range from near-sunlight to complete dark, and where swimming brings a buffeting stream of particles and small organisms. Thick mucus helps to protect the eye, while muscles can partly close the lid, or can retract, or perhaps even push out the eye. Internally, the rather thick eye lens lacks the ciliary muscles that, in other mammals, adjust lens shape and focus.

Because the eye is also focused for the dense medium of water, cetaceans out of water are quite short-sighted. Yet, captive animals clearly can see in the air, and may perform complex movements that require rapid visual and body coordination. Such behavior indicates better binocular vision than expected for animals with eyes on the side of the head. Wild animals sometimes "spyhop," looking about with the eyes generally directed ventrally (downward relative to body axis).

Underwater, vision seems to be important in shallow, sunlit waters. Variable color patterns in some species hint at a role for visual recognition of individuals. The eye of the Ganges River dolphin, which lives in muddy waters is very reduced, with no lens. It must rely on echolocation and touch for navigation and communication.

I SPY *Although probably short-sighted in air, the humpback whale (left) must be gathering information visually when it spyhops. A close-up of the eye of a gray whale calf (top).*

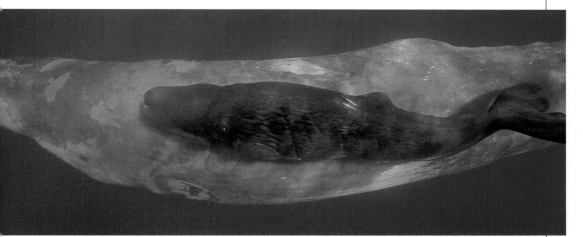

TOUCH *A sperm whale mother and calf (above) caress constantly. The tongue of the bottlenose dolphin (left) has taste buds and captive animals show preferences for particular foods.*

TOUCH

Whales and dolphins are very sensitive to touch. It is clearly important socially, because rubbing and sexual contact— with the same or opposite sex, and sometimes a different species—is common. Touch receptors, which may be densely packed in the skin, perhaps also function as pressure sensors in diving.

Cetaceans lack the prominent sensory whiskers of many other mammals, but scattered sensory hairs occur in many species. Young odontocetes often have a few coarse hairs along the upper lip, and adult mysticetes usually have well-developed lip hairs.

TASTE AND SMELL

Anatomical study of the tongue shows that taste buds are present. Captive dolphins can identify a modest range of experimental tastes. Also, captives may have quite particular food preferences, again hinting at a key role for taste. It is suspected that taste is important in recognizing the different body secretions used in social signaling, such as sexual receptivity or danger. A small sensory organ, Jacobson's organ, occurs at the front of the mouth in a few species. Perhaps this further broadens tasting abilities.

The sense of smell is a little contentious. Within the nose, mysticetes still have a small olfactory cavity, with scroll-like bones for nasal sensory tissues, and olfactory nerves, but the size points to a very reduced sense of smell compared with land mammals. Recent studies show that plankton produce a distinctive gas, dimethyl sulphide, which might allow mysticetes to

Seeing, hearing, feeling, are miracles, and each part and tag of me is a miracle.

"Song of Myself," *Leaves of Grass,* WALT WHITMAN (1819–92), American poet

track their prey by smell. The olfactory nerve has not been identified conclusively in odontocetes, and there is no separate olfactory cavity. Smell was presumably lost as the nasal passages evolved into the complex sound generators used in echolocation.

OTHER SENSES

The presence of pressure-compensating mechanisms in cetaceans points to the existence of appropriate sensors. During diving, for example, the middle ear cavity must be pressurized with air if hearing is still to function. Yet, details of these sensors are uncertain.

As in all mammals, the three tiny semicircular canals in the inner ear of cetaceans monitor acceleration and gravity. In cetaceans, the canals are smaller than might be expected, hinting at a reduced role in detecting rotation and acceleration.

The mineral magnetite (iron oxide) is a key biomagnetic material that occurs in many organisms, including some cetaceans. Biomagnetism allows animals to orient relative to the Earth's magnetic field. For cetaceans, biomagnetism may play a role in migrations and perhaps mass strandings. But, at the anatomical level, it's not clear how cetacean magnetite functions.

73

VOICES *from the* DEEP

All whales and dolphins deliberately make a wide range of sounds and use them in communication.

Cetaceans make good use of the properties of sound underwater. Sound can be better than sight for long-distance communication both under and above water. It travels nearly five times faster in water than in air, and low frequencies may travel very long distances indeed. Odontocetes also use the high frequency sounds they produce to echolocate (see p. 76).

VOCALIZATIONS

As a group, cetaceans produce sounds that range from very low to very high frequencies. Some individual species also have a wide frequency range, and most species produce several different styles of sound. Baleen whale sounds include rather low-frequency moans, belch-like sounds, bellows, snorts, bubble sounds, knocks, grunts, and yelps. Dolphins produce higher frequency

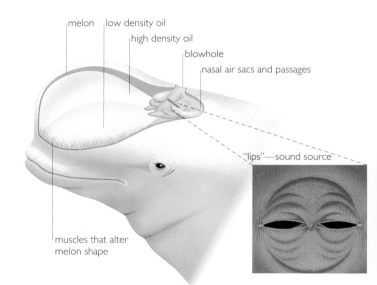

melon | low density oil
high density oil
blowhole
nasal air sacs and passages
"lips"—sound source
muscles that alter melon shape

SOUND PRODUCTION *We are still learning how sounds are produced by belugas, but the high-frequency sounds used in echolocation are probably focused in the melon.*

whistles, squeaks, and clicks, singly, or in bursts, or in a continuous stream. Dolphins can also make "jaw-clap" sounds in conflict situations. Other noises, like splashes and slaps made by the body, may also be used to communicate.

HOW SOUNDS ARE MADE AND HEARD

Mammals usually make sounds using the larynx, a complex structure in the airway to the lungs. Elastic vocal cords in the larynx vibrate as air passes across them. Despite years of study, though, it is still not clear how, or even if, cetaceans use the larynx to produce sound. A laryngeal source has been suggested for baleen whales, in the absence of other convincing sound generators, but there are no vocal cords in mysticetes. Perhaps other parts of the larynx help to produce sound. Mysticetes also lack the complex soft tissues near the blowhole

DEFENSE MODE *These Atlantic spotted dolphins are squawking to issue a threat.*

SOUNDS AND DIALECTS IN ORCAS

The orca (inset) produces complex "discrete calls"—distinctive sounds that may act as acoustic "identity badges" for individual pods. These calls, mostly pulsed, repetitive sounds that are stereotyped enough for humans to recognize basic patterns, are common during travel and foraging. This suggests that they are signals that help to coordinate activity. Each pod has a distinct and rather stable dialect, but calls vary slightly among individuals, perhaps because of the way each learns the calls when young. Orcas also produce more variable calls, used when the animals are close together resting or socializing.

which, in odontocetes, produce the high-frequency sounds used in echolocation. Vocal folds occur in odontocetes, but it is not clear if these produce the whistles reported for many species.

Sound is received in the ear by an amplification chain of three tiny ear bones that transmit sound to detectors in the inner ear. But how does sound get to the start of the amplification chain? The ear canal, which in humans and other mammals is open, is closed by a wax plug in mysticetes, and is vestigial in odontocetes. In odontocetes, sound travels from the side of the head to a thin "acoustic window" or "pan bone" in the lower jaw. From there, a fat body directs sound to large ear bones and then to the amplification chain.

How mysticetes hear is not clear, for although extinct species had a "pan bone" of sorts, living species lack this feature. In both odontocetes and mysticetes, the left and right ears are isolated from one another by complex air-filled sinuses inside the head. These sinuses probably help cetaceans to better detect the direction of the sound source.

LONG-DISTANCE CONTACT

Low-frequency sound travels long distances underwater. Baleen whale sounds have been detected from hundreds of miles away, and distances of thousands are possible. In the ocean, differences in salinity and temperature create extensive bodies of deep water which, because of their unusual density properties, may act as long-distance sound channels, transmitting low-frequency sounds even across entire oceans.

LANGUAGE STUDIES

Whatever we might want to believe about the acoustic behavior of whales and dolphins, years of study have not identified language in the human sense. In human language, variation in the order or context of familiar words produces widely varying meanings, and our words are associated with ongoing change in behavior. Cetaceans clearly produce complex sounds, but language-like transmission of information between individuals leading to change in behavior has not yet been demonstrated clearly. It could be that our monitoring techniques of sound or behavior are too crude.

HUMPBACK WHALE SONGS

Male humpback whales (inset) sing under the water, producing a complex series of low grunts, squeals, chirps, whistles, and wails. A song is made up of anything from two to nine separate themes, which are sung in a specific order, and can last from a few minutes to as long as half an hour.

Songs are produced mainly in tropical (winter) breeding grounds, but singing may also occur in the cold-water feeding grounds in summer and autumn. In any one ocean basin, whale populations sing the same distinctive song. Over time, each song changes progressively, but all the whales manage to keep up with the changes. The function of songs is not clear. Some researchers suggest they may form part of sexual behavior, others suggest that they may help to maintain social order.

ECHOLOCATION

Toothed whales and dolphins are skilled echolocators, using sound to navigate and to hunt prey.

An echolocating dolphin sends out an intense beam of sound, usually at high frequency. This bounces off an object and returns as an echo, helping the animal to determine its distance, position, and size.

Echolocating odontocetes produce close-spaced clicks of broad-spectrum sound. Clicks, usually at high to very high frequencies, vary in structure, with each species having distinct frequencies and patterns.

Echolocation is known for certain in only a dozen or so species studied in captivity. But all the wild odontocetes studied also produce similar sounds to captive animals. This points to echolocation as a key behavior pattern in all odontocetes—sperm whales, pygmy sperm whales, beaked whales, the Ganges and Indus river dolphins as well as other river dolphins, true dolphins, porpoises, narwhals, and belugas. Strangely, though, echolocation is not used all of the time by all species. Orcas, for example, echolocate to hunt fish, but don't always use it when hunting other cetaceans or seals. (Visual hunting is better to find prey which, because of their own good hearing ability, could detect an approaching predator.)

NO HIDING PLACE *A bottlenose dolphin (above) using echolocation to find tiny fish hiding in sand. Echolocation works by bouncing a beam of sound off objects (right) and evaluating the echoes.*

Also, it seems that there is a learned component to echolocation—it does not just come naturally to all animals.

HOW DOES IT WORK?
Odontocetes produce the sounds used in echolocation in the complex soft tissues of the nasal passages, inside the forehead between the skull and the blowhole. A large, fatty, internal structure, the melon (see diagram p. 74), apparently helps to transmit sound forward and out into the water as a narrow beam. In the sperm whale, the skull itself probably acts as an acoustic reflector to further refine the sound beam.

Theoretically, when sound is transmitted through different media, such as bone, tissue, and water, a loss of efficiency would be expected. It is not clear how odontocetes avoid "transmission loss" for outgoing and returning sound.

Echolocation requires complex identification of different sorts of sound. The target distance is determined from the length of time it

takes for an echo to return, but this means that outgoing sound must be distinguished from faint echoes. Also, the sensitive ear must be isolated from the intense noise of the sound-producing region. Isolation is provided by complex air-filled sinuses inside the head. These probably also separate the left and right ears from each other, allowing directional (left–right) hearing.

Returning sound is picked up at the side of the head and is transmitted to the ear via a succession of structures: the "pan-bone" in the jaw, a fat body, a large ear bone (which functions instead of an eardrum), and an amplification chain of three tiny bones (see p. 73). Within the ear, the cochlea, a spiraled organ, detects sound and passes acoustic signals to the brain. Just how the brain processes and interprets the signals is uncertain.

WHY IS IT SO IMPORTANT?
Echolocation works in the absence of light, and is effective in turbid waters both

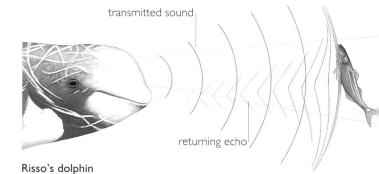

transmitted sound

returning echo

Risso's dolphin

near the shore and in rivers (where there is much suspended sediment), and also in dark, deep waters below about 150 feet (45 m). So, echolocation opens almost all waters of the world to the odontocetes. Judging from the form of fossil skulls and ear bones, odontocetes were able to echolocate from their earliest beginnings, about 30 million years ago. This complex behavior may help explain why odontocetes have diversified more in structure and lifestyle than any other group of marine mammals.

ECHOLOCATION, SONAR, AND RADAR

Odontocetes use sound exactly as humans use sonar, except that we've used sonar for about 50 years while they have been echolocating for millions. Sonar uses emitted sound and its returning echo to detect the presence, distance, and position of objects. Like odontocetes, bats also use this mechanism. Many other mammals (such as rodents, insectivores, and carnivores) hear high frequencies, but there is little clear evidence that they echolocate. Radar,

invented by humans, involves essentially the same principles of emitted energy and returning echo, but uses radio waves rather than sound.

HOW EFFICIENT IS ECHOLOCATION?

It was observations on captive odontocetes that led to the recognition of echolocation. In the 1950s, it was shown that blindfolded dolphins could swim easily among obstacles without collisions, and that dolphins could navigate well in the dark. Experiments have shown that some species can echolocate well enough to detect targets about 1/16–1/8 inch (1.5–3 mm) in diameter, from distances up to 10 feet (3 m). Some dolphins reportedly can discriminate between ball bearings 2 inches (5 cm) in diameter and those of 2.5 inches (6.5 cm). Generally, it seems that odontocete sonar is a short- to intermediate-distance sense,

FINE FOCUS *An Atlantic spotted dolphin has no trouble catching a tiny needlefish.*

APPROPRIATE MEANS *While the two orcas (above) may track prey, especially fish, by echolocation, they will spyhop to decide on a strategy for the capture of these emperor penguins.*

which operates best up to about 30 feet (10 m).

DO BALEEN WHALES USE SONAR?

High-frequency sounds have been recorded occasionally from baleen whales, such as minke and blue whales. These sounds are a minor component of the wide range of noise produced by mysticetes, and they lack the structured form produced by echolocating odontocetes. Mysticetes also lack the complex nasal passages which, in odontocetes, generate echolocation sounds. It seems, therefore, that mysticetes do not use sonar in any major way.

WHALE MIGRATION

The annual seasonal movement of whales between

feeding and breeding areas is known as their migratory cycle.

The seasons drive the migrations of all creatures, including a number of whales. Seasonal effects are greatest in high latitudes, where long summer days and melting of sea ice bring about massive blooms of microscopic plant plankton, or phytoplankton. These are eaten by krill, copepods, and other zooplankton, which in turn feed birds, seals, squid, fish, and whales. When polar seas freeze over in winter, biological production slows and many species migrate to warmer climes, to waters comparatively devoid of marine life.

BALEEN WHALES

Baleen whales are the greatest cetacean rovers, spending almost as much time traveling as they do on their breeding and feeding grounds. Except for Bryde's whales, which remain in warm waters year-round, all baleen whales make essentially north–south migrations between cold-water feeding areas for summer, and temperate to tropical breeding areas for winter.

Pacific gray whales show a typical pattern. In early winter, perhaps in response to various hormonal changes initiated by shortening days, they move south to breed in the warm, shallow lagoons along the Mexican coast. From February, they migrate north to feed along the coast of Alaska and in the Beaufort Sea. Mothers

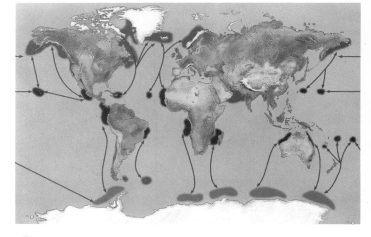

● breeding areas
● feeding areas

NO BOUNDARIES *This map shows the migratory patterns of the different populations of humpback whales. The fin whale (above left), is a fast and powerful migratory traveler.*

with newborn calves remain in the breeding areas longer than the others, to allow their calves to gain strength for the long swim ahead.

Humpback whales show a similar pattern. In the North Atlantic, they breed around the West Indies, migrating up the western side of the Atlantic Ocean in spring, to disperse in summer feeding areas between the Gulf of Maine and Iceland. A smaller population migrates between Norway and west Africa and the Cape Verde islands. North Pacific humpbacks breed along the coast of Mexico, around the Hawaiian Islands and southern islands of Japan, feeding along the coasts of the northern Pacific Rim.

The seasons are reversed in the Southern Hemisphere, so humpbacks breed between June and October along the tropical coasts of the southern

continents, and around Pacific islands such as New Caledonia and Tonga. These populations summer in the krill-rich Southern Ocean surrounding Antarctica. The greatest confirmed migration by any individual mammal is that of the humpbacks that summer on the Antarctic Peninsula, south of Cape Horn, and breed off the coasts of Colombia and Costa Rica. In the Arabian Sea, one unusual population of humpbacks does not migrate, because its warm breeding waters support abundant marine life for food.

Most southern right whales remain in the mid-Southern Ocean, but some feed at the edge of the Antarctic pack ice. Their coastal breeding grounds lie mainly along the southern coasts of Africa, South America, and Australia. The breeding destinations of blue, fin, sei, and minke

STEADY PACE *Although humpbacks (right) can move at 5 miles (8 km) per hour, they rest and socialize often, and average about 1 mile (1.6 km) per hour overall on their epic journeys. A migrating gray whale yearling calf (below) plays in a kelp forest off the Californian coast.*

whales are almost entirely unknown, but probably lie in warm, deep, tropical waters.

TOOTHED WHALES

Most toothed whales do not have the well-delineated migrations typical of baleen whales. Many are nomadic rather than migratory. Orcas were long thought to migrate away from Antarctic waters, but at least some of them breed and winter in the ice. One species whose migration is understood is the sperm whale, which has a unique pattern. While female sperm whales and immature males remain in temperate to tropical waters during summer, mature males migrate to polar waters to feed on vast quantities of squid, rejoining the nursery schools in winter to breed.

WHY MIGRATE?

In terms of energy, migration is a huge commitment. Baleen whales are believed to fast throughout much of their migration, living on their blubber and fat for up to eight months of the year. In the case of a female that calves on migration, total weight loss may be as high as 50 percent. In a fully grown blue whale cow, this may represent a loss of some 89 tons (81 tonnes). Migrating whales probably have a "tight energy budget."

The conventional view is that birth and early development in warm water is important for calves, and that insufficient food remains in the feeding areas during winter. But bowheads, orcas, belugas, and narwhal raise their young in very cold water, and plankton supports huge numbers of Antarctic seals, penguins, fish—and many whales—during the winter.

It is possible that before continental drift, migratory endpoints were much closer, and that whales have simply continued to visit these points as they drifted farther apart. Perhaps baleen whales breed away from Antarctic waters to avoid predation on their calves by orcas, which do not migrate. Some animals not involved in breeding forgo migration, saving a huge expenditure of energy. Many minke whales winter in the Antarctic sea ice, and about half of the Southern Hemisphere adult female humpbacks in any year do not reach the breeding grounds. Instead, they may remain in feeding waters year-round, putting on condition for the next year's cycle.

Large rorquals, such as blue and fin whales, are the most powerful travelers, with one fin whale recorded as averaging 10½ miles (17 km) per hour over 2,300 miles (3,700 km).

UNIQUE PATTERN *Of the sperm whales (below), only the adult male migrates long distances.*

... *As they fell down sideways they splashed the water high up, and the sound reverberated like a distant broadside.*

Journal of a Voyage round the World
CHARLES DARWIN (1809–82), English naturalist

CHAPTER FOUR
SOCIAL LIFE *and*
BEHAVIOR

LIVING TOGETHER

The social organization of cetaceans is often a consequence of the ecology

of their prey, and the degree of cooperation required to catch it.

One of the most fascinating things about watching cetaceans is observing their social lives. There are three overriding priorities: feeding, reproduction, and avoiding predation. Mothers and calves form the basic unit of society common to all cetacean species, but beyond this, differences start to emerge.

Feeding may be the single most important factor in determining the nature of cetacean societies. The single greatest division is between baleen whales and toothed whales. One group has baleen plates, used to swallow large numbers of small schooling prey; the other has teeth, to grasp single, larger prey.

TOOTHED WHALES

Toothed whales occupy a vast diversity of aquatic habitats, from the deep polar oceans to equatorial rivers and estuaries. Their group size varies from almost solitary, to great

aggregations of thousands of animals. Although toothed whale groups are usually more stable and larger than those of baleen whales, species such as river dolphins are found in very small groups, or even singly. They live in a stable environment in which their prey is more or less evenly dispersed, and they have to contend with fewer predators.

Dolphins that live along coastlines face more predators, and their food is more clumped, so they tend to be found in larger groups. The largest cetacean groups are those of oceanic dolphins, which travel widely in search of scattered patches of prey. They move in small, closely

related, permanent groups, which may amalgamate with similar groups to form temporary herds of thousands.

The close relationships within toothed whale societies give them their stability. Most species are female-centered: that is, adult females form the stable nucleus of the group. Males leave at puberty to join other groups, thus avoiding inbreeding. There are rare exceptions to this, such as orcas and pilot whales, in which all members remain with one group for life, and mating occurs between two such groups. This leads to a very high degree of social cohesiveness and cooperation, well demonstrated in coordinated hunting by orcas.

Another well-known and often tragic illustration of the cohesiveness of toothed whale groups is the phenomenon of mass strandings (see p. 98). Science cannot yet comment on the personal relationships among individual whales, but dry genetic theory predicts that it is worth the risk to help related companions if your shared genes then have a better chance of surviving.

Social life, however, is not

GROUP OF TWO *This pair of humpbacks (left) may be foraging or merely enjoying each other's company.*

INDEPENDENT *Some toothed whales, such as the Amazon river dolphin (left), are often found alone or in small groups.*

lives of other rorquals. Large aggregations are sometimes seen where there is abundant food. In Antarctic waters, one group of nearly 50 fin whales and another of 25 blue whales were sighted. There was also a report of nearly 1,000 minke whales feeding on a huge swarm of krill. Yet almost nothing is known of social interactions within these species. Blue whales have been reduced to such low numbers, particularly in the Southern Hemisphere, that contact and socializing may be quite difficult for them.

In breeding areas, once again, most of what we know comes from humpback and southern right whales (see pp. 96 and 147). On breeding grounds, humpback social groupings are usually small and unstable. Groups of up to eight humpback males compete aggressively to escort females prior to mating, but there do not seem to be long-term associations among males as there are with right whales. After weaning, juveniles tend to be solitary, becoming more social with age.

It's them as take

advantage that get

advantage i' this world.

Adam Bede,
GEORGE ELIOT
(1819–80), English novelist

all sweetness and light. Hierarchies of dominance exist, in which subordinate animals are forcefully reminded of their place. These are most clearly seen in captive dolphins that cannot avoid each other. In the wild, dominance may be used to settle access to food, or to mates.

Body language and underwater sounds express social status, but conflict often becomes physical. Tooth rake marks are very common in many toothed species, and indicate the frequency of aggressive interactions in everyday life. In some beaked whales, males have a single pair of teeth used for display or fighting, and many animals are covered with linear scars.

BALEEN WHALES
Much of what we know about baleen whales comes from the two most easily observed species, humpbacks

and southern right whales. Social organization in feeding areas derives from the nature of their food source—usually scattered patches of small fish or crustaceans, better utilized by small groups. In such diverse humpback feeding areas as the Gulf of Maine and the Antarctic, the most common group size is two. Where their gender has been determined, they are either females, or a male and a female. It has been found in recent years that these pairs may be together over several years, suggesting that these animals form good foraging "teams." It may also be, of course, that like dogs, horses, and humans, they simply prefer the company of certain whales over others. Mature males do not associate in feeding areas, perhaps unable to put aside the intense rivalry of the breeding season.

Little is known of the social

MATING *to* BIRTH

There is nothing more important in the lives of cetaceans than the perpetuation of their own genes, and hence their species.

Animals have an inbuilt compulsion to repro- duce. Each of the 81 cetacean species has evolved a unique combination of timing, behavioral strategies, and physiological adaptations to ensure that it survives.

Compared to many other mammals, cetaceans have low reproductive rates. This is because they grow slowly and do not mature sexually for at least five or six years (much longer than some other species), and almost invariably give birth to only a single calf, which takes a year or more to reach independence.

This low reproductive rate is offset by the fact that they are generally long-lived, with females capable of bearing many calves during their life.

INTIMATE MOMENT
Atlantic spotted dolphins mating (with male underneath).

BREEDING CYCLE

Scientific study of the ovaries and testes of dead whales and dolphins over many years has resulted in an understanding of the animals' breeding condition at various seasons. From these, it appears that reproduction is a cyclical activity. In many species, the breeding cycle is intimately linked to the seasons and, in the case of migratory species, to migration.

This is particularly so in baleen whales, which have a surge in hormonal activity in both sexes as they approach the breeding areas, possibly stimulated by changes in day length or water temperature. The exception is the tropical Bryde's whale, which breeds throughout the year.

Some oceanic dolphin species have two breeding peaks—in spring and in fall. Other species, such as coastal bottlenose dolphins, may breed throughout the year, as their environment changes little with the seasons.

This linked breeding and migratory cycle means that the breeding season is short and intense. The gestation, or

BEGINNINGS OF LIFE *The diagrams (right) show the male and female reproductive organs of a typical dolphin.*

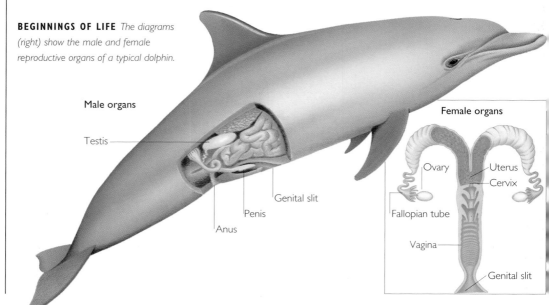

Male organs

Testis

Genital slit

Penis

Anus

Female organs

Ovary

Uterus

Cervix

Fallopian tube

Vagina

Genital slit

Birth, and copulation, and death. / That's all the facts when you come to brass tacks.

Sweeney Agonistes,
T. S. ELIOT (1888–1965),
American-born English poet

pregnancy period, in many species is usually around 12 months, so mating occurs on the calving grounds. The female will return a year later to deliver her calf.

Huge energy demands on breeding females, including fasting during migration, mean that only very rarely do they breed in consecutive years (although a humpback cow has been seen with new calves in three consecutive years). More often, there is a breeding cycle of two or three years in baleen whales. In many toothed whales, the cycle is even longer, because gestation may be longer (up to 18 months in sperm whales) and calves stay with their mothers for longer.

MATING

Once males and females arrive at the breeding areas, complex mating systems come into play. No species of cetacean is

monogamous: they are either promiscuous (males and females have multiple partners, as is common among right and gray whales and many dolphins), or polygynous (males mate with multiple females, as seen among sperm whales, orcas, and narwhals).

The songs of polygynous male humpbacks advertise a smorgasbord of choice for females. Intense competition among males often follows, as they attempt to displace each other to become a female's "principal escort" as a prelude to the gentler acts of courtship and mating.

Males of other species employ a variety of strategies, from the mating alliances of bottlenose dolphins (see p. 92) to narwhals, which use their spectacular tusks in jousts for supremacy, sometimes with fatal results. In right whales, sperm competition (see p. 96) is a more peaceful solution. Males, not necessarily related, may actually help one another to mate, then let their sperm fight it out on the way to the ovum (although they are not above raking each other with barnacle-covered callosities). Larger penises are seen in

CONTEST *As two Atlantic spotted dolphins are mating, a rival male blows bubbles in a bid to cut in.*

animals that employ sperm competition, as extra length may give a competitive edge.

Females are not mere passive recipients of all this male activity. Females have their own priorities, which are to mate with a suitable male, and to raise their offspring successfully. While mature males are usually ready to mate, females may be in a non-breeding phase of their cycle, or may exercise their right to choose a mate.

It is difficult for a male cetacean to force a female to mate against her will. Females may signal their sexual status in various ways, perhaps by means of hormonal secretions in the urine, or by subtle behavioral cues that are obvious to prospective mates.

PREGNANCY

In migratory species, such as humpback and gray whales, newly pregnant females are the first to leave the breeding areas and return to summer feeding grounds. A pregnant female must rapidly put on enough fat to sustain her and her growing fetus through the coming year. This is a difficult task, bearing in mind that most baleen whales fast during migration. Near-term cows are among the first to return to the calving area, as early calving allows more time for the calf to grow before it, too, must migrate.

BIRTH *to* DEATH

Although their lives are perilous from birth, many

cetaceans have a life expectancy comparable to that of humans.

Except in captivity, the birth of cetaceans has rarely been witnessed. It is a dangerous time, both for the mother, who may be distracted by her labors, and for the calf, which relies on its mother (and possibly "aunts," or helpers). While captive dolphins have been seen to bear their young tail-first underwater, a gray whale calf in Baja California emerged head-first, above the surface. For cetaceans, single births are the norm, with the calf being typically a third the length of the mother.

Once the calf has emerged, the mother guides it to the surface for its momentous first breath. This is the start of a close bond which in some species will persist for life. In others it is severed in a year.

Once mating has taken place, the father's role is finished, although adult males may defend closely related calves or those that have a high chance of being their offspring. In species such as sperm and pilot whales, defense of the young is a communal responsibility.

EARLY LIFE

For the first few weeks of life, a calf rarely strays from its mother's side, and the two are in frequent physical contact. Small, weak, and defenseless, calves cannot dive deep or swim far, and require constant feeding in order to grow to a size where they can begin to fend for themselves. As they grow, they become bolder and venture farther from the safety of their mother or their group. Feeding bouts become longer, but fewer. Growing calves, like other young animals, can be insistent about being fed, and southern right whale cows, at least, have been seen to refuse to feed their calves, provoking what can only be called tantrums.

Cetacean milk is very rich, consisting of about 40 percent fat. The lactation period varies considerably. Humpbacks feed their calves for nearly 11 months, while minke, sei, and fin whales lactate for between 4 and 6 months. Blue whales lactate for seven months, during which the calf will grow at a rate of about 180 pounds (85 kg) per day to nearly 25 tons (23 tonnes).

All baleen whales gradually wean their young after the first year of life. Southern right whales appear with their calves at the breeding area exactly 12 months after giving birth. They simply leave their calves and depart. At this point, the calves become

NURTURE *A humpback (top left) and her calf. A sperm whale calf (left) will nurse for several years, in some cases up to 15 years. Part of the umbilical cord can be seen still attached to the belly of the newborn sperm whale (below).*

Life is a gamble at terrible

odds—if it was a bet,

you wouldn't take it.

Rosencrantz and Guildenstern are Dead,
TOM STOPPARD
(1937–), English playwright

HOW LONG DO CETACEANS LIVE?

A general rule in nature is that smaller animals live shorter lives. Harbor porpoises, one of the smallest cetacean species, live only 15 years or less, while it seems that some large whales may survive for 100 years. Bottlenose dolphins live for about 50 years in the wild. Until now, age could be determined only by examining a cross-section of a toothed whale's tooth, or a cross-section of the waxy earplug that forms inside a baleen whale's ear. Both of these are built up in layers, each of which corresponds to a yearly cycle. Layers also form in the very thin band of cement just beneath the outer enamel of the tooth, allowing for the most accurate calculations. Now, with the photographic identification of natural markings, it is possible to study the lives of individual whales from birth to death, without harming them.

Cross-section of a toothed whale's tooth showing growth layers

Cross-section of the waxy earplug showing growth layers

juveniles, and appear to have little further contact with their mother. This may be the basis of the apparently loose social ties in baleen whale society. In contrast, most toothed whales feed their calves for 2 years or more, with some old sperm whale cows still nursing calves that are 10 to 15 years old.

After a young cetacean has left its mother, it must make its way in a dangerous world. Mortality is highest during the first years of life, as disease, predators, parasites, pollution, injuries, deaths in fishing gear, vessel strike, and hunting take their toll. There are some un-expected causes of death and injury, such as a dolphin being suffocated by an octopus cov-ering its blowhole, and whales have been found with billfish spears embedded deeply inside them. Mortality rates are difficult to calculate, but it is believed that a majority of whales reach adulthood.

GROWING UP

As cetaceans grow, their play behaviors eventually become the serious activities of life. Playful contests between young males become the mating battles of adults, and animals such as orcas actively teach their young the foraging and hunting techniques that are essential for survival.

Alliances that are important for mating and feeding success are made. When cetaceans near sexual maturity, their lives change. Baleen whales generally mature early, at around five to seven years old, with sexual maturity being related to body size. In depleted populations (such as Antarctic sei whales) individuals have increased access to food, so growth is faster, and sexual maturity is achieved earlier. In

toothed whales, young males that have lived within their mother's group reach puberty and most leave to join groups of other young males. In sperm whales, full sexual maturity is correlated with size, with the larger males not maturing until the age of 20 or more, and smaller females at 10 years.

CETACEAN INTELLIGENCE

Many cetaceans have relatively large and complex brains, but does this make them "intelligent?"

Even in humans, "intelligence" is a loaded term. It can mean many things—the ability to know, to analyze using reason and judgment, to think fast, and to understand. But what does variation in human intelligence mean? And how can we identify intelligence in other species?

INTELLIGENCE QUOTIENT

Human intelligence is related to context. How we measure someone's intelligence is highly dependent on setting. Assessments of human intelligence—IQ tests—are standardized to allow easy comparison of results, so it is ironic that standardization brings loss of context.

Beyond behavioral testing, structural features of the nervous system have been used to gauge intelligence between species. Many studies of mammals have looked at the cerebral hemispheres—the largest parts of the brain. In those lobes, a well-developed outer layer (cortex) seems to indicate higher intelligence. Another measurement, the encephalization quotient, an index of brain size to body size, also seems to be linked with intelligence. Species with a bigger brain than expected for their body size appear to be more intelligent.

There are many anecdotes about cetacean intelligence which involve, for example, belief in interspecies communication between human and dolphin, and even extrasensory perception (ESP). Can we get beyond this level of anecdote in our understanding of cetacean intelligence? The serious cetologist must go farther.

In general, cetacean brains are larger than expected for their body size (relative to other mammals) so that the encephalization quotient is high. Their large brain size points to a need for information processing which is related to complex hearing, communication, and movement. This relative brain size, though, is not always high: small river dolphins and the large sperm whale have smaller than expected brains.

Behavioral tests have been used extensively for cetaceans, but because cetaceans are non-manipulative, and don't modify their environment in the way that most mammals do, it is not clear what these tests may mean.

ABILITY TO LEARN

Studies on a few captive species of odontocetes show that they can quickly learn complex physical routines. Captives learn to respond to

LEARNED BEHAVIOR *The orca (top) is playing with a young sea lion. Often, a pup is stunned or killed in this way. The learning ability of the dolphin (left) is studied at a dolphin research center.*

CARE-GIVING BEHAVIOR *An adult short-finned pilot whale (right), together with some other members of the pod, closely guard a dead calf.*

different human gestures or sounds, modifying their behavior in response to subtle changes in the instruction given. There is some evidence that learned information is transmitted from one animal to another. But human-like language, with its implications for intelligence, has not been identified clearly. For wild animals, there is strong evidence that odontocetes all have the ability to echolocate, but there is also a strong learned component in successful echolocation.

PROBLEM SOLVING

The activity of cetaceans in the wild shows that many species use very sophisticated problem-solving behaviors. Cooperative bubblenet feeding in humpback whales (see p. 66), and beach capture of seals by orcas, are examples of complex monitoring and physical coordination. Wild cetaceans show care-giving (epimeletic) behavior that may involve supporting a calf or sick animal (see p. 71). Some species show clear group defensive behavior (see p. 71). So, cetaceans produce complex responses to rapidly changing environments. If these behaviors are learned, they indicate high intelligence, but if they are genetic (instinctive), they are not such persuasive evidence.

HOW WHALES COMPARE WITH OTHER ANIMALS

Encephalization quotients are high for most cetaceans. And, fossils show that cetacean brains became large quite early on in evolutionary history, long before the increase in human brain size. But brain size alone has to be interpreted carefully, because other animals with absolutely and relatively small brains can show extremely complex behaviors. Pigeons, for example, have small brains, yet complex flight patterns. Among mammals, bats have tiny brains, yet they use highly sophisticated echolocation to detect prey and determine complex flight paths. Perhaps a key feature of many cetaceans is that, compared to many mammals, they have a complex and human-like folding of the cortex in the cerebral hemispheres. Parts of the brain that process sound are also well developed.

HOW HUMAN-LIKE IS CETACEAN INTELLIGENCE?

What can we make of similarly large and complex brains? In humans, brain complexity and intelligence are apparently linked with sophisticated language and with manipulation of the environment. Cetaceans don't modify their environment in the human sense or, indeed, as other mammals do, and a human-like language has not been demonstrated. Brain complexity perhaps relates to sound processing. It is not clear why brains should be similarly complex in odontocetes and mysticetes, which produce and use different sorts of sound. We have far to go before we can really understand if cetaceans are intelligent and what their intelligence might mean.

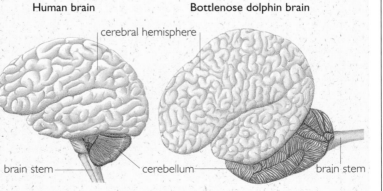

Human brain

Bottlenose dolphin brain

cerebral hemisphere

brain stem — cerebellum — brain stem

INTERPRETING BEHAVIOR

One of the most mystifying things about cetaceans is trying to work out just what they are doing while you are watching them.

Behaviors vary from the dramatically obvious, to the almost invisibly subtle. While it takes much study to gain a comprehensive knowledge of the behavior of a species (and we are far from achieving this for any species), it is possible to learn something of the types of cetacean behaviors and so make some sense of our observations.

LEAPING AND PORPOISING

Probably the most popular image of cetaceans is that of a leaping, porpoising, or breaching animal. These are all behaviors in which the whole body, or most of it, breaks clear of the water. Smaller animals, such as dolphins, generally leap and this may signal a general state of excitement, or perhaps something more specific. Off Argentina, dusky dolphins leap to show that they have found aggregations of fish and that more dolphins are needed to herd them so all the group may have a feed. Many species leap, but none so spectacularly as the spinner dolphin. Their spinning leaps are thought to coordinate and synchronize behavior within groups.

Although superficially similar to leaping, porpoising is not a social behavior but an energetically efficient method of travel. Small cetaceans clear the water when surfacing to breathe, as air offers less resistance to movement than water. Nevertheless, a high degree of synchronization is often seen among porpoising animals, and it is an eye-catching sight when large schools are on the move.

BREACHING

One of the grandest sights in nature is a breaching whale. This is the term given to the leaping of cetaceans. The animal seems to burst from the water in slow motion. Humpbacks are probably the most prodigious breachers, and individuals have been seen to breach more than 100 times consecutively. A number of explanations have been proposed, ranging from simple *joie de vivre*, through the dislodging of parasites (such as barnacles and remoras), to getting a better look at boats, or warning away intruders. All of these may be true, but breaching is certainly a visible and audible form of communication, although what is being communicated is difficult to know. In one case, a humpback calf that had been separated from

MYSTERIOUS URGE *Pec-waving (top) is a common activity of humpbacks. A breaching humpback (above) may repeat its leap many times.*

its mother breached for at least two hours. Breaching may follow disturbance by boats, or by other whales. Sometimes it seems contagious, and mothers and calves, or other pairs of cetaceans, often breach in unison.

HEADRISES AND SPYHOPPING

Other behaviors where the head breaks the surface include headrises, in which the animal's head lunges briefly out of the water while moving forward, and spyhopping, a similar behavior, but usually while the whale remains stationary. Whales often spyhop to scan their surroundings, as visibility in air is usually much greater than underwater. Whales may spyhop to inspect boats, and orcas in Antarctica spyhop to examine seals and penguins resting on ice floes.

FINS AND FLUKES

A different set of behaviors involves waving or slapping the pectoral fins or the tail flukes. "Pec-slapping" is commonly seen in humpback whales, which lie on their sides, slapping their long pectorals against the water surface, to produce a resounding crack that can be heard at a considerable distance. Pec-slapping can also be used more gently between courting animals close to one another. Right whales and many other species use their pectoral fins for reassurance, or to caress one another during courtship.

Tail flukes have functions other than propulsion. They are defensive weapons, and right whales, for example, have only to point their flukes at another animal to send a clear warning to stay away. Many species slap their flukes on the water, or "lobtail," as it is called. As with breaching and pec-slapping, this sends a visible and audible signal, with a variety of meanings.

It may be a leisurely, almost lazy, activity used by socializing whales, or it may be a forceful threat. In extreme examples, open-boat whalers often had their boats smashed by lobtailing whales. Southern right whales also use their flukes in another way—called "sailing." They hold their flukes vertically out of the water for extended periods. Whether they are actually sailing, cooling themselves in the breeze, or simply enjoying the sensation of the wind, is not known.

WHAT'S GOING ON? *Humpbacks often indulge in boisterous pec-slapping (left). The spyhopper (below) in this pod of socializing melon-headed whales, may simply be wanting a better view.*

CHAMPIONS *Spinner dolphins (above) spin their bodies rapidly around their long axis while they are in the air.*

BOW RIDING

Another familiar behavior is bow riding, when dolphins, in particular, ride, almost without effort, in the pressure wave created ahead of the bow of a vessel or the head of a whale. Larger whales, such as orcas, also bow ride. When observing bow-riding dolphins, you will often notice other behaviors signifying aggression or sexual activity. These include biting, butting, exerting dominant status to drive another animal out of the bow wave, and overt sexual activity, including copulation.

BOTTLENOSE DOLPHINS

The best known cetacean species is the bottlenose dolphin.
Because it lives along coastlines in many parts of the world,
it is the most often seen and best studied cetacean species in the wild.

Studies, some of them extending over more than 20 years, of coastal bottlenose dolphin societies have enabled a detailed understanding of their behavior and social organization. The lives of their offshore relatives, which live in the open sea, are much less well known.

SOCIAL UNITS

A long-term study at Sarasota Bay, Florida, by Dr. Randall Wells and colleagues, identified four types of social units: a mother-calf pair, groups of subadults of either sex, adult females with their calves, and adult males. The way these groups interact is largely determined by the breeding biology of the species. In this "fission-fusion" society, individuals constantly leave and rejoin groups.

MALE BONDING

Bottlenose dolphins are slow breeders; females calve only every four to five years, so females in mating condition are rare in a population at any given time. As a result, males have developed strategies to make the most of mating opportunities when they occur.

Males leave their mother's group at puberty and join groups of subadults. As they age, they form smaller and smaller groups, until fully

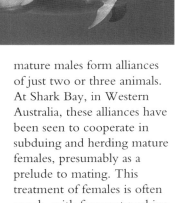

CLOSE CONTACT *The bottlenose dolphins (above and top) socialize and play in a variety of ways. Mother and calf (left) will maintain a long association.*

mature males form alliances of just two or three animals. At Shark Bay, in Western Australia, these alliances have been seen to cooperate in subduing and herding mature females, presumably as a prelude to mating. This treatment of females is often rough, with frequent pushing, buffeting, biting, and vocal threats, and females are often unwilling participants. Males sometimes cooperate to herd females; at other times they compete, and even fight for females. There appear to be social hierarchies, often dominated by the largest males, and much time is spent reaffirming bonds. The pecking order can be seen when they are feeding on discarded fish from trawlers— some stay aside to let others get the best discards.

COOPERATIVE EFFORT *Bottlenose dolphins driving fish up onto a river bank.*

FEMALE NETWORKS

While male alliances appear to be among the most lasting bonds in bottlenose society, some female associations also last for many years. Suckling of calves can continue for four years or more, and some female calves may accompany their mothers for life. Other calves join subadult groups with a preponderance of males, in which social behavior is explored and developed.

Many adult females form loose, changing networks. In broad biological terms, males devote their energies to impregnating as many females as possible, while females devote their energies to the raising of offspring. There is no breeding advantage to females in forming permanent close relationships, but they can call on their associates when assistance is required for feeding, defense, or in some cases, repelling amorous males. In some cases, "aunts" will assist a female in giving birth.

Dolphins frequently stroke and caress each other, and sexual behavior, including copulation, is used casually, apparently indiscriminately, to reaffirm bonds. Aggressive contact is also common, with the scars from tooth raking seen on a majority of animals. Eavesdropping with hydrophones reveals that angry squawks, clicks, and pops accompany aggressive behavior.

Bottlenose dolphins also show a decidedly unsocial phenomenon—the occasional solitary animal that is never seen in company with other dolphins, and which lives a life apart from the security offered by a group. Occasionally these loners associate with humans instead, perhaps to satisfy a need for company. We do not know if they have left voluntarily or have been driven out.

COOPERATION

Cooperative behavior is regularly seen in feeding bottlenose dolphins. Indeed, they have a remarkable ability to adapt to a wide variety of feeding situations, and have devised clever ways to catch prey. In South Carolina, they are known to cooperate to drive fish up onto a muddy river bank, then beach themselves briefly to grasp the flapping fish.

While animals that spend their lives together are likely to exhibit a high degree of coordination, cooperative feeding does not necessarily imply altruism—by agreeing to cooperate, all benefit as individuals. When searching areas with less prey, bottlenose dolphins often forage singly or in small groups.

Some populations or communities of bottlenose dolphins reside in bays or estuaries where there is a constant food supply. They sometimes forage outside in open water, but generally remain in a restricted area—their "home range." Within a "community home range," individuals may have preferred foraging areas. Along coastlines, small populations may live in overlapping home ranges, which they patrol in search of food, traveling in groups for defense. One such community comprises about 100 Sarasota Bay dolphins. Mating probably occurs among such adjacent groups.

TACTILE CONTACT *Social relationships are constantly reinforced with touching, from gentle caresses to more violent pushing.*

SPERM WHALES

*An animal of extraordinary appearance,
the sperm whale has equally extraordinary habits.*

Herman Melville, in his classic story *Moby Dick*, declared the sperm whale king of whales, and expressed the fascination seafarers, and whalers in particular, felt for this archetypal whale. The largest of the toothed whales, the sperm whale is the deepest of all divers, possibly descending as far as 10,000 feet (3,000 m) in its search for squid. It is found in all seas from the equator to the edge of the polar ice.

EARLY PERCEPTIONS

The perception of the social organization of sperm whales was for many years based on aspects of human society. Mature males are much bigger than females, and it appeared that sperm whale society was dominated by powerful males who maintained harems of females and their young. These harem masters ruled by defeating challengers in combat, leaving the females only for the annual summer migration of males to feed in polar or subpolar waters.

These perceptions fitted the observable facts at the time, when sperm whales were mainly encountered by whaling

MUTUAL GOOD *Sperm whales in the North Atlantic (right and below) appear to have different levels of association, perhaps moving in and out of groups in response to the amount of food present.*

expeditions that had no interest in patiently observing whale behavior. Only in the past two decades has the approach to cetacean research changed sufficiently to enable the truth to be revealed. What is emerging is a complex picture that will continue to unfold for many years to come. In essence, the lives of mature males and females are quite different.

NURSERY SCHOOLS

At the center of sperm whale society is the so-called "nursery school," comprised of adult females and immature animals of both sexes, including suckling calves. These schools are found near oceanic islands, seamounts, and continental shelves in equatorial to temperate regions, rarely straying north or south of 45 degrees latitude. Off the Galápagos Islands, a research team led by Hal Whitehead found three levels of association within nursery schools. Small groups, called "units," average 13 animals and remain together for years and possibly for life. These units often join into "groups," which average 23 animals and remain

DISTINCTIVE SHAPE *The blunt-nosed sperm whale (left) is the largest of the toothed whales, growing to more than 60 feet (18 m).*

In 1983, Dr. Hal Whitehead and several companions sailed into the Indian Ocean in the tiny 33 foot (10 m) sloop *Tulip*, to begin a program of sperm whale research in the Indian Ocean Whale Sanctuary—particularly in the area off the east coast of Sri Lanka. This was the first time that such research had been attempted. Tracking groups of whales day and night by their underwater clicks, they used the small, self-sufficient vessel to locate and stay with the whales for periods of days and weeks. They identified and got to know individuals and gained unique insights into their social organization, behavior, and ecology.

Geographical remoteness and increasing civil strife made it impracticable to continue the research in Sri Lankan waters, so in 1985 Hal (pictured below), his wife Lindy, and their colleagues continued their research around the Galápagos Islands in the equatorial east Pacific. There they continue to record and unravel the complex social relationships and communication of these mysterious whales, and to throw new light on their feeding and diving behavior, and ecology. With Hal Whitehead's entry into the field, the study of sperm whales entered a new phase.

together for about a week. Several groups may then join into "aggregations," which form for only hours, and average 43 whales. It is likely that the more abundant the food, the larger the groupings.

There is a high degree of social cooperation within these nursery schools. It is obvious that a nursing mother who must dive deep for up to half an hour or more to feed, needs help to protect her calf, which must remain near the surface. Within permanent units, the care of calves is a shared responsibility. Females born into such a unit will probably remain with it for life, developing strong ties to its other members. Young males, on the other hand, apparently leave, or are made to leave, on reaching puberty between 7 to 10 years of age.

BACHELOR SCHOOLS

Young males then join other socially immature males in "bachelor schools," which live separate lives from the nursery schools, although they may occur in the same area. As the males mature, they continue to associate with males of a similar age, and form smaller bachelor schools. Finally, when they become physically mature at around 27 years of age, other males become competitors rather than allies, and from that point, males are more often solitary. Indeed, fully grown bulls have come to be regarded as almost always solitary. They make very distinctive underwater vocalizations that may serve to keep them apart. However, the presence of broken teeth and jaws, and extensive scarring on their heads indicates that fighting does occur among mature males.

It seems that males compete for access to the reproductive females in a nursery school, and only one male at a time will mate with the females in a school. He will usually stay with such a school for only a matter of hours, before moving on to find another school. Yet in whaling days, mature males often defended nursery schools that they were attending: the sinking of the whaler *Essex* in 1820 seems to have been an example of this.

FASCINATING STUDY *A female southern right whale (above) and calves at the Head of Bight, South Australia. Roger Payne and Katherine Payne (below) watching a southern right whale in the Gulf of San José, Argentina.*

SOUTHERN RIGHT WHALES

Southern right whales have revealed a social system of surprisingly subtle complexity.

In 1970, the study of southern right whale society was initiated by Roger Payne and his then wife Katherine at Península Valdés, Argentina, where it had been noticed that these whales gathered every year. This was exciting news, because right whales, both southern and northern, had been comprehensively slaughtered by whalers during the nineteenth century, and even after their protection in 1935, their recovery had been hampered by illegal poaching.

For the same reason that they were once the easiest whales to kill, southern right whales are the easiest species of cetacean to study—during their breeding season, they congregate very close to shore, often in shallow bays. As a result, they can be easily observed from headlands and cliffs in areas such as the Gulf of San José at Península Valdés without being disturbed. This is a mating and calving area, and Roger and Katherine soon became aware that a rich tapestry of social life was unfolding before them.

MATING
Like many other species, right whale females mate with a succession of males. Males move from female to female, seeking those that are sexually receptive. Unlike humpbacks and some other whales, there is little violent competition between males for reproductive access to females; indeed, right whales are generally most peaceable. Instead of physical competition, right whales engage in "sperm competition." Males have the largest testes in the animal kingdom, each pair weighing about one ton and producing copious quantities of sperm.

It is thought that actual competition occurs among sperm (see p. 85), with males waiting patiently for each other to finish, provided the female is willing. There seem to be alliances among males to help each other by restricting a female's movements so that they can mate more easily. If a female, larger than the male, doesn't want to mate, it is difficult to force her to do so. She may move into very shallow water to dissuade her suitors, turn on her back, or simply swim away from them.

SOCIAL CUSTOMS
It was long thought that right whales were basically solitary, and did not form lasting relationships. But females seem to form preferred groups with other females, at least within seasons. Whale researcher Stephen Burnell has studied southern rights at a major calving site at Head of Bight,

South Australia, since 1991. Mothers tend to keep newborn calves apart from other whales, but as they grow and begin to explore their environment, they are allowed to interact with other whales, usually mothers and calves.

Not all mothers and calves are acceptable: some are threatened if they approach, others ignored, while favored ones may approach. It is not uncommon to see a female with two calves swimming beside her, while the mother of one of the calves patiently waits, submerged, nearby. Occasionally, a calf will leave with the wrong mother, and then its own mother must turn and rush back to it. Meanwhile, small groups of subadults may prowl like teenagers in a shopping mall, looking for action.

PHYSICAL COMMUNICATION

Right whales are extremely tactile animals, and subtle in their behavior. While whales such as humpbacks are very vocal and probably mediate much of their behavior by the use of underwater sound, right whales spend a good deal of their time in physical contact

EASY VIEWING *A group of three subadult whales (above) visit a female and calves at Head of Bight, South Australia. Close-up (right) of a southern right whale breaching.*

with others, although socializing adults also vocalize. A right whale will signal its displeasure at being approached by another whose company it does not want, not by some extravagant display, but simply by orienting its body so that its flukes, its main weapon, point at the interloper. This message is picked up at a considerable distance, and usually provokes a change in course away from the unwelcoming individual. Courting adults make surprisingly sensuous use of the pectoral fins, stroking and caressing one another.

Mothers and very young calves are in almost constant touch, and as the calves grow and explore farther away, they frequently return for reassurance. This is partly related to their feeding: young calves feed frequently, but in short bursts, while older calves feed less often, but for longer. The gentleness of a cow with a

boisterous calf has to be seen to be believed. Since the calf depends on the cow's milk supply, and this is not infinite, discipline is sometimes needed.

Yet, despite the closeness between mother and calf during the first few months of life, the following season when they return to the breeding area, the mother promptly abandons her well-grown offspring, and returns to the Southern Ocean feeding grounds, leaving the yearling to make its own way in society. As yet, there is no evidence of a special relationship between mothers and calves in later years. Newly weaned yearlings soon learn to associate with others, perhaps including calves they met during their first season.

97

THE MYSTERY *of* STRANDINGS

Of all the varied aspects of cetaceans' lives, none has such

a grip on the public imagination as strandings.

Usually well covered by the media, the sight of a large group of whales or dolphins dying on a beach, or being pounded by surf, provokes feelings of pity, incomprehension, and impotence. How do such superbly adapted marine animals come to be out of their element and helpless?

TYPES OF STRANDINGS

Animals may be alive or dead when stranded; they may be alone or in groups, sometimes very large. The great majority of whales, dolphins, and porpoises die out at sea, and are either consumed by sharks while still afloat, or their bodies may sink and be eaten by the small animals of the sea floor. The bodies of stranded dead animals have probably drifted ashore with winds or currents while still inflated with the gases of decomposition. These are usually single animals, unless there is a fatal epidemic afoot within a local population. Sometimes dead strandings give vital clues to the cause of death, which may include net entanglement, ship strike, or toxic pollution.

WHY DO THEY STRAND?

Humans have long been perplexed by live strandings. Prehistoric humans would certainly have used stranded cetaceans for food and other

HIGH AND DRY *Beached false killer whales at Seal Rocks, on the east coast of Australia in 1992.*

useful products, and are likely to have had their own explanations of why these animals had come ashore. The Greek philosopher and natural historian Aristotle, among the first to write of the phenomenon, had the sense to admit that whales strand "for no apparent reason."

More recently, speculation has abounded, and for many years cetacean "suicide" was a popular theory. Strandings have even been linked to extraterrestrial life and cosmic events. Another theory proposes that the animals are trying to return to their evolutionary roots on land.

More sensible explanations have come from serious studies, which have revealed that the species that strand most often, and in greatest numbers, are those that form cohesive social groups and are commonly found in deep

water away from coastlines. These include pilot whales, false killer whales, and sperm whales. Their unfamiliarity with the coastline may be a factor. Species that spend much time along coastlines, such as humpback and southern right whales, rarely strand.

Deep-water animals may follow prey inshore; indeed, sperm whales, considered to be the deepest diving whale, have recently been found feeding in very shallow water off New York. Other correlations suggest that whales have navigational "malfunctions." Serious infections of parasitic worms are sometimes found in the ear canals or the brain, which may affect the whales' coordination, orientation, balance, and hearing. Weakness

RARE EVENTS *A single juvenile humpback (left) in Western Australia, and more than 400 long-finned pilot whales (below) stranded in Catherine Bay, New Zealand.*

or confusion caused by injury or disease may drive animals into the shallows, where life can be prolonged a little without drowning.

Coastal topography seems to play a crucial part. Many strandings occur where there are gently sloping beaches. In such places, perhaps the echo-location signals of toothed whales are deflected away by the sloping bottom, indicating open water ahead.

In New Zealand, headlands of a certain configuration are thought to shepherd whales into dangerous waters, particularly when adjacent to a shelving bottom. Sandbars shift with time, perhaps confusing even animals that are familiar with an area. Weather may also play an important part in many strandings: onshore winds and heavy seas could hamper cetaceans retreating from a dangerous situation in shallow water, especially on a falling tide.

Another possibility is that cetaceans may navigate using anomalies in the Earth's magnetic field. Migratory birds and marine turtles, for example, have this ability, and magnetite, a mineral involved in geomagnetic navigation in other animals, has been found in the brains of some cetaceans. Such a capability could help to explain the whale's marvelous ability to navigate over great tracts of apparently featureless ocean.

Yet such a sense could also lead cetaceans astray. In Britain, a correlation has been noted between strandings and locations where magnetic lines of force cross the coast at right angles, although another study in New Zealand has shown no such correlation.

The worst stranding on record was a herd of more than 400 long-finned pilot whales in New Zealand in 1985, but there have been many others involving scores or hundreds of animals. The key may be in the strong social bonds among members of many toothed whale societies. If an individual is sick, injured, or dying of old age, the lifetime habit of assisting group members in trouble is not broken because of proximity to shore.

This may seem noble or foolish, but such social structures are the key to survival for these species. This was not understood for many years, and the notion that they were bent on suicide was reinforced because rescued animals often restranded immediately. In reality, they were showing an extreme reluctance to leave their fellows, possibly even their mothers or calves.

IN THE SHALLOWS *The slope of the beach at Seal Rocks, Australia, may have disoriented these false killer whales.*

DEALING *with a* STRANDING

Surveying the scene of a mass stranding, or of

the stranding of very large whales, can be a heartbreaking experience.

E ven in good weather, with a calm sea, it may seem that little can be done to help. With large whales, this is usually true. Their sheer bulk usually renders them immovable, and the weight of their own tissues, when unsupported by water, may suffocate them or cause fatal damage.

Strandings of dead animals present a major disposal problem. Decomposing whales can constitute a potentially serious public–health risk, and efforts are usually made to remove and bury them in marked sites, so that the bones can later be exhumed for scientific collections. With large whales, this removal may be difficult. In several cases, attempts have been made to dispose of large whale carcasses with explosives—with spectacularly unpleasant results.

Living whales and dolphins face multiple stresses and present much more complex difficulties. Out of their

ALL HANDS ON DECK *Volunteers keep a pod of 65 stranded long-finned pilot whales (above) wet and protected from the sun until the tide turns.*

HELP FOR THE HELPLESS GIANTS

I f you are the first at the scene of a stranding, what can you do? Someone should notify the police or wildlife authorities as soon as possible. Often, the first instinct is to try immediately to return the animal to the sea as shown below. Such attempts may injure it, or end with it restranding. Until expert help arrives, here are some first-aid measures that you can take to try to stabilize the whale's condition:

🐋 Never pull on the head, dorsal fin, flippers, or flukes, but try to roll the animal upright, so that its blowhole is clear of the water or sand.

🐋 Try to orient its body up the beach, away from the breaking waves. Avoid getting too close to the flukes and the teeth, because when distressed, the animal may thrash about and injure or bite you.

🐋 Rinse the eyes and blowhole clear of sand, taking great care to avoid pouring water or washing sand into the blowhole when it is open.

🐋 Keep the skin from drying out, either by pouring water over it or, preferably, by covering the animal with a cloth and keeping the cloth damp. Bring buckets and sheets or blankets, if available.

🐋 Remain calm, and talk to and stroke the whale quietly and reassuringly. Unless spectators are prepared to help, ask them to keep at a distance, and to remain quiet.

🐋 As long as they are made comfortable, and are not bothered by excessive noise or crowds, whales and dolphins can rest comfortably for some time, until other measures can be put in place. They seem to appreciate efforts made to help them.

KEEPING AFLOAT *Helpers (above) maneuver a stranded short-finned pilot whale onto a floating mat for transport to a nearby rescue facility. False killer whales (right) are prevented from restranding in Augusta, Western Australia.*

element, they are often highly distressed, not least because of separation from their pod members. They may be at the mercy of the surf, pounded on rocks, or rolled upside down, their blowholes and eyes filled with sand. Suffocation by sand or drowning are common causes of death in stranded animals. If they are lucky, they may refloat themselves on the next tide—belugas frequently do this. But often they are beached—dried and burned by sun and wind, sometimes on remote beaches that humans can reach only with difficulty.

Remembering that toothed whales accompany a sick or injured companion into a stranding situation, it is very useful if the first animal ashore can be noted, and marked with a flag or something similar. Rescue attempts can focus first on this animal, or a veterinarian may decide to euthanase it humanely, if it is beyond hope of a successful return to the water. It may

then be easier to rescue the other animals.

Single animals are usually returned to the sea or in some cases taken to a rehabilitation facility to be treated. If they don't succumb to the original injury or disease that caused the stranding, they are returned to the water when they recover. Mass strandings require the logistics of a military operation and the help of many people, using techniques that have been perfected during many strandings over a period of years.

On some occasions, this involves removing the animals from the scene by truck or trailer, and taking them to a sheltered waterway nearby (if one is available), where they are reunited with their surviving companions and stabilized before release into the ocean again. Pontoons may be used to float and move animals. Sometimes there is little that can be done, but there have been some

amazing successes with these methods. Near Augusta in Western Australia in 1986, 96 of 114 false killer whales were successfully returned to the ocean over three days.

For biologists, strandings provide snapshots of the lives of the species involved, including rare species. You may see researchers measuring and photographing live animals, and dissecting those that have already died. This yields valuable data about the animals' diet, their physical and reproductive condition, and other factors, such as parasites or pollution, which may have contributed to their deaths. Some people find this activity ghoulish and upsetting, but researchers are involved because they care about the animals. Much of the information that contributes to the management and conservation of these animals comes only from strandings.

There's nothing like a jaw-to-jaw encounter with a friendly shark or whale to get the juices running.

Sylvia A. Earle (b. 1936), American marine biologist and conservationist

CHAPTER FIVE

IN *the* FIELD

GETTING *into the* FIELD

Once you have decided to take your interest in whales to the practical

stage of going out to look for them in the wild, how do you go about it?

The world of cetaceans is vast and varied, and options for viewing them depend on what you can afford in time and money and, to a certain extent, on where you live. A resident of Monterey, California, or of Sydney, Australia, will find it easier to go whale watching than someone living inland.

The first step is to find out where and when whales, dolphins, and porpoises can be found. The activities of many species are influenced by the seasons, and there is a wealth of information concerning the time and place different species can be found. Watching Whales, Dolphins, and Porpoises (page 194) features 30 locations where you are most likely to encounter cetaceans. Like migrant birds, the movements of whales cross national boundaries, so it is possible to see the same population of, for example, gray whales feeding in the Gulf of Alaska in July, and then breeding along the Pacific coast of Baja

AERIAL VIEW *Two gray whales, one blowing, spotted from a plane window.*

California, Mexico, in February. In many areas, cetaceans are present year-round. Some areas are famous for sightings of one species; others host many.

BACKGROUND DATA

Take the time to do some research, according to your interests. Perhaps there is a coastal area that you have long wished to visit, and you may be able to combine your visit with whale or dolphin watching. On the other hand, if you are interested in a particular species, you could plan your holiday around their movements.

While there are many whale-watching sites of world importance and reputation, successful encounters with cetaceans are

possible at many lesser known areas. Local knowledge can be obtained from natural history museums or the National Marine Fisheries Service in the United States (or the equivalent government wild-life authorities elsewhere). You could also consult the commercial whale-watching industry or local residents of particular areas. Check out the information available on the Internet. Extend your knowledge by joining a volunteer group that organizes whale-watching activities, including counts and surveys of scientific value. Some of these are listed in the Resources Directory (page 272).

TYPES OF ENCOUNTER

What sort of whale-watching experience do you want? Would you like to experience the brief, close encounters offered by commercial whale-watching vessels? Would you prefer the more leisurely, less intrusive approach of observing coastal migrations from the shore? Or are you likely to take your own vessel into waters frequented by cetaceans? Going whale watching by yourself gives you flexibility of timing and movement. On the other

MARVELOUS CREATURES *Atlantic spotted dolphins enjoy playing with this diver who is using an underwater scooter.*

Blue seas ... are the domain of the largest brain ever created, / With a fifty-million-year-old smile.

Whale Nation,
HEATHCOTE WILLIAMS
(1941–), English poet

hand, placing yourself in the hands of others may give you access to whales you otherwise cannot have, as well as to the knowledge and expertise provided by professionals.

Remember that watching whales and dolphins usually requires patience. If you are short of time, by all means join a scheduled cruise, and take your chances—you may or may not see the animals. The more time you spend watching for cetaceans, the more varied and interesting behavior you will see.

The range of whale-watching experiences is very broad. You can wade with bottlenose dolphins at Monkey Mia in Western Australia, observe humpbacks from inflatable boats on the Antarctic Peninsula, or river dolphins from a wooden boat on the Amazon. You can watch the passing of migrating gray whales from Californian headlands, or feeding fin whales in the Gulf of St. Lawrence in Canada. You can watch humpback whales glide over the warm shallows of Silver Bank in the Dominican Republic, or sperm whales diving off the South Island of New Zealand.

A rise in some baleen whale populations since whaling ceased, coupled with the increase of global tourism and commercial whale watching, means that many more people are now encountering

NEW FRIENDS
A tourist feeding the dolphins at Monkey Mia, Shark Bay, Western Australia.

cetaceans than even a decade ago. Backed by unprecedented levels of public interest, this has increased awareness of and concern for dolphins and whales, and has had considerable economic benefits for many coastal communities. Numbers of recreational boaters have also grown sharply, and they often encounter whales. Never before has there been such scope for the cetacean enthusiast.

PREPARING *for a* FIELD TRIP

*Like any outdoor activity, whale, dolphin, and porpoise watching
requires careful preparation, for comfort, safety, and enjoyment.*

Once you have decided where and when you wish to look for whales, there are several things to consider. Will you be on the water, or on land? Are you going to be exposed to the elements for long periods of time? Will you be in a remote area?

If you are going to sea and you don't know if you are prone to seasickness, you should assume that you are. Consult your physician or pharmacist for an approved remedy, such as acupressure patches or oral medication.

PROTECTION

Remember that when you board a commercial whale-watching vessel, you should be shown what safety equipment is on board, and where it is stowed.

Whale watching in all its forms involves some exposure to the elements: wind, sun, rain, heat, cold … even sleet and snow. It is essential that you are prepared for any conditions you might meet. Not only will being caught unprepared (so that you become sunburnt, wet, or cold) spoil your experience, it may even be dangerous.

So think about where you're going and what climatic conditions you can expect—not only the average, but the extremes. You may be surprised. Tropical regions, for example, do not always exhibit the "balmy breezes" of the tourist brochures. Often the trade winds blow at almost gale force, sometimes for weeks at a spell. Desert regions, such as Baja California in Mexico, can be witheringly hot and windy, while mild temperate regions, such as southern Australia, can experience sudden cold changes, even in summer. It is best to be over-equipped with warm and dry clothing if possible—one of your companions may need extra protection one day.

Windproof wet-weather gear and a good sunscreen are probably

BE PREPARED *You never know what the weather might do so always have warm, waterproof clothing and protection against the sun and glare off the water.*

the most important items, particularly if you are on the water. When dressing for cold conditions, use the "layer principle"—several relatively light layers of clothing, which trap insulating air between them. In this way, when temperatures change, a layer or two can be donned or shed. Hypothermia and sunstroke are killers, and with current levels of ozone depletion, ultraviolet radiation is a serious health risk. Deep tans no longer make sense, but sunscreens do. Good sunglasses, either polarized or UV-proof, will protect your eyes from the intense glare on water, and also make it easier to see whales in bright conditions. Remember that about 35 percent of body heat is lost from the head, so take a warm hat or cap, whatever the conditions.

PERSONAL COMFORT

The inner person also needs looking after during long sessions searching for cetaceans. Snack food is always welcome, and cold or hot drinks may be essential in extremes of climate. Each adult will need about 2 pints (1 L) of fluid per day, considerably more in hot, dry conditions. A vacuum flask is a good investment for both hot and cold drinks or soups.

You may wish to prepare for periods of boredom, when

Blue Whale   Humpback Whale

whales are not to be seen.
You can read (a good time to
read up about cetaceans),
sketch or paint, watch birds,
or simply watch and listen to
the sea in its changing moods.

WHALE-WATCHING GEAR
The basic equipment needed
for watching cetaceans is quite
simple: something to see them
with, and information to help
identify what you see.

Most important of all are
binoculars, the eyes of any
observer of wildlife, and they
should always be close at
hand. A good pair of
binoculars will convert that
distant splash into a living
animal. They will enable you
to identify whales at a

DISTANT VIEWS
*A spotting
telescope (left) is
great for land-
based
viewing.*

distance, and to follow details
of their behavior and the
subtle features of their
appearance that would be
invisible to the naked eye. If
you can afford to buy a good
pair of binoculars, they are
well worth the extra expense.
Buying cheap ones may be
false economy, as their poorer
optics give a lower quality
image, and tire the eyes more.

You should select a pair of
binoculars between 7x and
10x magnification—the
number of times closer an
object appears through the
binoculars than with the
naked eye. Anything less than
7x is not powerful enough,
and any more than 10x is
probably too heavy, and hard
to hold steady. You will
notice when looking at
binoculars that two numerals
are given, for example, 7x50,
or 8x40. This second numeral
(50 or 40) refers to the
diameter of the front lens in
millimeters (1 inch = 25 mm).
The greater this number, the
more light the lens lets in and
the more effective it is in low-
light conditions. But this also
results in heavier binoculars.

You have to find a balance of
weight, viewing clarity, and
power that suits you. More
expensive binoculars, of the
roof-prism type, tend to be
more compact and lighter for
a given magnification, while
porro-prism ones are bulkier
and heavier. Waterproof
binoculars are more costly.

Spotting telescopes are
useful, especially for land-
based watching and on cruise
ships, but are too powerful
(generally between 15x and
60x) to keep steady on small
boats. From land, to watch a
distant whale, they are ideal,
but they can't replace binoculars, and are usually used after
a sighting has been made with
binoculars or the naked eye.

You should also have at
hand a good field guide to
cetaceans, such as this book,
or perhaps one that specializes
in the area where you are.
Watching whales becomes
much more interesting when
you know which species you
are observing. You can then
put your own observations
into a framework of what is
known about the species, such
as its migrations and behavior.

RECORDING OBSERVATIONS

Watching whales and dolphins is usually an experience that stays in the memory, but you may wish to keep a more tangible record of your sightings.

Apart from photography and video, there are a number of ways you can record events: in your personal diary, keeping more detailed observations in a journal, through to keeping a computer database of your sightings. Or you may prefer to sketch or paint your subject.

KEEPING A JOURNAL

Not only is this an excellent way of remembering your sightings, it is also a very useful discipline, encouraging you to remember details, at least until you can commit them to paper. Your journal should be of a size that can be stored in a day-pack. Try to find one with robust binding and, if possible, waterproof pages. The humble pencil (HB or softer) is the simplest and most reliable writing implement in the outdoors—as long as you have a sharpener or a penknife at hand.

The skill of keeping accurate records comes with experience, but there are basic items of information which, if regularly gathered, will not only add to your experience and to your store of

LOGBOOKS
Keeping good records and sketches of sightings can make a valuable contribution to whale research.

knowledge, but may prove useful to cetacean researchers. The trick to studying these animals is in being at the right place at the right time, and there may be times when you will be the only person at the scene, perhaps witnessing something that has never been recorded before.

WHAT TO RECORD?

Every recorded sighting should be put in the context of where (either a location or a geographical position) and when (date and time) it occurred. Note what the weather and sea conditions

were at the time, and if there were any other obvious features that may be associated with the sighting, such as schools of fish on the surface, large numbers of seabirds, the presence of boat traffic, or predators, such as sharks.

Often you get only a brief glimpse of part of a whale— they don't make it easy for you. Training yourself to look for the presence or absence of certain features may help. For example:
● Are the animals large, medium-sized, or small? These are the simplest subjective categories of size.

WHAT WHALE WAS THAT? *Two researchers consult their previous records to help make a positive identification.*

- Try to estimate their size relative to your boat, to their distance from the shore, or to any other useful scale.
- Is there a dorsal fin? Is it prominent or small? Does it have a distinctive shape? Are the pectoral fins short or long?
- Are there any distinctive markings, patterns, or scars on the body or tail flukes? What colors are the whales?
- Is the top of the head flat, with visible bumps, knobs, or ridges? Are there prominent whitish growths around the head? Is there a protruding beak and a bulging "melon?"
- If the mouth opens, are teeth visible? Or baleen?
- Is the blow strong or weak? Bushy or a tall column? Vertical or angled forward? Does it form an obvious "V?"
- How many animals are in the group? Are they close together or spread out? Do there appear to be any calves?

It is very important not to make assumptions about what you see. During the hump-back migration, for example, you shouldn't assume that every whale you see off the coast is a humpback. Other species may migrate through the same waters at the same time. It is best simply to describe what you see, for example, "three large whales; strong, bushy blows; obvious dorsal fins; dark color." This may be all you have time to notice before they disappear.

Many species can easily be confused with similar species. On the other hand, if you see a knobby forehead, a very long pectoral fin, or serrated flukes with white undersides, you are justified in identifying a humpback whale. Similarly, if you spot a "large to very large whale, with a

RECORDING BEHAVIOR

This may give clues to important activities, such as feeding or breeding as well as to the whale's identity. Some examples:

Are they stationary? Are they moving quickly or slowly? Are they traveling steadily, or milling around, meandering, or stopping? Are they close to the shore, to reefs, or to floating ice, or to your boat or another boat?

Are they submerging for seconds, minutes, or tens of minutes? If you have a timer on your watch, the timing of dives and other behavior intervals can be extremely informative.

Are they behaving quietly, or energetically? Are there obvious displays, such as breaching, spyhopping (as the baby orca, right, is doing), lobtailing or pec-slapping? Are they moving fast in groups, with much splashing and apparent excitement?

Are they associated with large numbers of seabirds, seals, and so on? Are they obviously feeding? Can you catch a sample of their prey, perhaps with a net or bucket?

completely smooth back, no dorsal fin, and white growths on the head," it is undoubtedly a right whale.

Using a hand-held cassette recorder to note details of complex whale-watching events is highly recommended. Rather than attempting to re-member what you are seeing, or trying to commit it to paper as it happens, simply record a commentary and transcribe notes from it later.

For identification purposes, sketching any remembered features in a small notebook is preferable to using a tape recorder. For a few minutes after a sighting, important details may stay vividly in the mind, including some surprisingly minute ones that may be very useful later on. Don't worry too much about

YOUR CHOICE *Observations can be kept in anything from a notebook (above) to a laptop computer (right).*

the quality of the sketch, just try to put pencil to paper as soon as you can, while your impressions are still strong.

Cetacean behavior is notoriously difficult to interpret, and there is much that is still unknown, so don't be worried if you cannot determine what the animals are doing. Simply try to describe what you see—you may be able to make sense of it later, or talk it over with someone who can help you.

If you are really keeping up with the times, you may wish to record your sightings on a laptop computer. There are software programs available for keeping wildlife databases. You should always back up your files as soon as possible.

PHOTOGRAPHING WHALES

Few subjects can be as satisfying—or as difficult—to photograph as cetaceans.

CLOSE-UPS *An orca surfaces in front of a lime-kiln lighthouse (left). Watchers observe a humpback whale (below).*

Everyone who has photographed whales longs for a brilliant shot. A photograph of a whale breaching, for example, is the result of a combination of many things: opportunity, timing, knowledge of how cameras and films interact, perhaps good balance, and a bit of luck. An understanding of the animal's behavior allows you to predict, up to a point, what they might do.

The most useful factor is the right camera gear. While poor photographs can be taken with good equipment, it is much harder to take good pictures with a poor camera.

PHOTOGRAPHIC GEAR

Although there is an increasing variety of compact cameras, many with zoom lenses, the most suitable type for whale photography is the single lens reflex (SLR). This camera is relatively compact, and you can change the lens to suit the situation. Alternatively, there are zoom lenses with a wide range of focal lengths, from wide angle to telephoto. A motor drive is a useful option when the action is fast and furious, as it often is when whales or dolphins erupt into bouts of activity.

The focal length of a lens indicates its magnification power. A standard 50 mm lens, for example, portrays the scene as the eye beholds it. A 200 mm telephoto lens gives 4x magnification, a 300 mm lens gives 6x, and so on. Telephoto lenses are essential for most whale and dolphin photography because usually you cannot get very close.

When cetaceans closely approach boats, a standard lens, or even a wide-angle lens may be suitable. If you are limited to one lens, a 200 mm telephoto is probably the best bet for photographing whales. Protect your lens with a filter, which can fill a dual role by reducing glare.

The longer the focal length of a lens, the more light it requires. However, lenses are also rated by a second characteristic—the f-stop number (the amount of light that a lens admits). The lower the f-stop number, the more efficiently the lens admits light. So a lens rated at f2.8 will shoot at a faster shutter speed in given light conditions than an f4 lens. The "faster" lenses cost more.

Film also has speeds: the higher the ISO/ASA number, the less light it requires for exposure. But the faster the film is, the grainier the photograph. As a rule, use ⅟₅₀₀ second shutter speed to freeze movement, but ⅟₂₅₀ is often adequate. A balance of shutter speed (for sharpness of image), focal length, f-stop, and film speed will come with practice.

Film of ASA/ISO 200 is usually adequate for use with a 200–300 mm lens at these speeds, and is not very grainy. More expensive (lower f-stop) lenses can use slower film, and the resulting photos will have finer image resolution.

FROM LAND

Cetacean photography from shore is like other wildlife photography, having to contend with the vagaries of weather, and requiring patience, the right equipment, and preparedness for the moment. While you are limited in your approach to your

subject by the shoreline, cetaceans can often be seen from land. You will almost certainly need a 200 mm telephoto lens or longer. Lenses of up to 1,000 mm are used, but long lenses are heavy and hard to hold steady; with lenses longer than 300 mm, a tripod is essential. If the animals spend much time in a locality, you may wish to walk along the cliffs to keep close to them; or you may prefer to remain comfortably in one place where you can watch them, waiting for interesting behavior.

Think about the weather and the time of day. Be careful changing lenses where dust or sand is blowing. A strong wind can overturn an un-attended camera tripod, with expensive results. Rain can also ruin camera equipment.

Early morning and late afternoon sunlight can be best for photography—but not if you are shooting into glare. Underexposed shots into glare with a whale silhouetted can be dramatic, but overexposure rarely brings good results.

> *'Tis said, fantastic ocean doth enfold / The likeness of whate'er on land is seen.*

Fish Women: On Landing at Calais, WILLIAM WORDSWORTH (1770–1850), English poet

FROM BOATS AND SHIPS

Photographing cetaceans from on the water introduces a new range of difficulties, and these usually increase with smaller vessels. While the deck of a ship can be as stable as land, the motion of small boats can be violently unpredictable. A good rule among seadogs is "a hand for yourself and a hand for the boat"—in this case a hand for your camera and one for the boat. In other words, while moving about, always hang on with one hand and do not entirely trust to your balance. Find a spot where you can wedge your lower body securely, leaving your upper body and both hands

free to weave with the boat's motion and deal with the camera. It is difficult initially, but try to keep the horizon level in your viewfinder—it can make a big difference to the quality of a photograph. Shutter speed should be $\frac{1}{500}$ second or faster, because the combined motion of the boat and the whale usually make sharp photos impossible with slower speeds.

Salt water is the arch enemy of camera equipment; spray can ruin electronics with surprising ease. In wet conditions, protect your camera (inside the front of your jacket, for example) until you are ready to use it. If it gets wet, wipe it with fresh water as soon as possible—it does much less damage than salt. Alternatively, special rain jackets for your camera can be purchased from most camera stores (or you can make your own). Change film where there is no risk of any water entering the camera back.

VIDEO

High-quality video cameras are now a serious alternative to still photography. Most have zoom lenses and work well even in very low light conditions—better than most still cameras. Video images can obviously capture the movement of dolphins, whales, or porpoises in a way that stills cannot. Remember that video equipment is even more sensitive to salt water damage than still cameras.

WHALE WATCHING
from the SHORE

There's a world of whale watching to be had from land, which in its own way offers as much as vessel-based experiences.

OFFSHORE *Whale watchers line the cliff top (left) to see gray whales off the west coast of North America each year.*

In the minds of many people, the archetypal image of whale watching is that of close encounters from the deck of a whale-watching vessel. But when you consider the vast extent of the Earth's coastlines, and how often many species can be found close to land, the potential for shore-based watching becomes obvious. With a little knowledge of where to go and what to expect, visits to coastal regions may take on the added excitement and satisfaction of watching cetaceans as they move and live along the coast.

Shore-based watching is free, you can do it independently any time you like, you can take your time, and combine it with your travels. You can do it in weather too rough for whale-watching vessels (although whales are difficult to spot in rough conditions), and you can enjoy the often beautiful and wild coastal scenery. Most important, you will not affect the behavior of the animals, because you are not intruding upon their environment.

The main disadvantage of shore-based watching is that you must depend on the animals coming within view-ing range. It is frustrating, for example, to see the splashes of a school of dolphins or the blows of a very large whale, off near the horizon, without any hope of being able to identify them. You must also be prepared to be patient and often to see nothing.

WHERE TO GO
Many scientific studies take advantage of the ease with which cetaceans can be ob-served from land. Migratory species, such as humpbacks and gray whales, tend to swim within 5 or 6 miles (about 10 km) of land, occasionally coming within a stone's throw while passing the Pacific coast of the United States. The cliffs of Point Reyes, California, provide spectacular views of

migrating whales, as do many of the headlands of northern California. In the sheltered waterways of Washington State, British Columbia, and Alaska, humpbacks, orcas, and other species may be easily observed from shore. Much farther north, bowhead whales stream around Point Barrow, Alaska, on their migration into the Beaufort Sea. The east coast of the United States offers less scope for shore-based whale watching. In the Southern Hemisphere, humpbacks are seen daily during their migration past Cape Byron, Australia's easternmost point, also well known for its surfing bottlenose dolphins. Off the arid Nullarbor coast of

MIGRATIONS *The movement patterns of many species are well known. Southern rights (left), swim close to shore. A mother and calf (above).*

South Australia, southern right whales congregate at the base of limestone cliffs for months during the breeding season, making this arguably the best whale-watching site in the world. They mate and rear their calves virtually at the watcher's feet, oblivious to the attention they are receiving. The Gulf of San José in Patagonia, Argentina, offers similar chances to observe right whales from shore, as does the southern coast of South Africa.

Cetaceans often enter fiords and bays to feed or rest. Bottlenose dolphins are the most commonly sighted coastal cetacean worldwide, because of their habit of coming close inshore. They are often seen in harbors, creeks, and estuaries. Some may become habituated to human contact, as has occurred at Monkey Mia, in Western Australia, where interaction

between humans and dolphins has reached such levels that it has been necessary to regulate the contact. Similar situations have developed elsewhere, and in some places, humans even swim with dolphins.

PLANNING ENCOUNTERS
There are two ways to plan your expedition. Either choose an area you wish to visit, or select a species of particular interest. Obviously your choices are somewhat limited, as not all species closely approach land on a regular basis. Many species, such as sperm whales, are found in deep water and are usually observed from boats. Your options will also be dictated by the seasons, and by the seasonal movements of the cetaceans themselves.

Obviously, coastlines with high cliffs or headlands will offer the best prospects for viewing cetaceans. When you are viewing from a cliff top, the horizon is many times farther away than when you are viewing from sea level,

JOINING THE FUN *Windsurfers find a novel way (left) to see humpback whales close to the shoreline in Bonavista Bay, Newfoundland, Canada.*

and many a pleasant hour can be spent tracking the distant blows of whales as they move along a coast. Binoculars or a spotting scope are a must, and you will develop the knack of finding a comfortable position where you can remain for hours. Wind and sun can take their toll, so find a sheltered spot if possible, and be prepared with hat and sunscreen. You may be able to camp at a suitable lookout: nothing is more enjoyable than the first sighting of the morning as you sip your coffee or tea.

With migrating species, such as gray and humpback whales, face the direction from which the whales are likely to come, and be patient. Occasionally scan around, looking for anything out of the ordinary—a fleeting dark shape, a splash, or the puff of a whale blow. Blows may be visible for 5 miles (8 km) or more through binoculars.

During migrations, there may be a constant stream of whales all day, with several pods visible at once. Try to follow individual pods, remembering that whales may submerge for 10 minutes or more. When waiting for whales to resurface, look ahead of where they dived— they usually maintain the same speed, whether at the surface or submerged.

WHALE WATCHING
from BOATS

Vessel-based whale watching takes you into the whale's domain, and offers experiences of an immediacy that cannot usually be gained from land.

The phrase "vessel-based whale watching" includes a wide range of experiences—from paddling a kayak among whales, to spending an afternoon on a commercial whale-watching boat, to viewing them from a ferry, or the bridge of a luxury cruise liner during an ocean voyage. These experiences fall into two categories—take yourself, or let someone else take you there.

ON YOUR OWN BOAT

Almost any boat can be used for whale, dolphin, or porpoise watching. The suitability of certain types of craft depends on the location, weather, water conditions, the operator's experience, and an awareness of how to behave around marine mammals (see p. 118). In sheltered waterways, small, fragile craft, such as sea kayaks or rowing boats are sometimes used to approach cetaceans. Only fully proficient users of such craft should share the water with whales. Although usually considerate toward small

BRAVING THE COLD (right) in Zodiacs to watch humpback whales in the pack ice of the Antarctic Peninsula. A dusky dolphin (above) riding the bow wave of a small boat, Kaikoura, South Island, New Zealand.

boats, when whales are engaged in vigorous activities, such as feeding or social interactions, they may easily bump or swamp a small craft that has drifted too close, and this is no time to learn self-righting techniques.

Some vessels have less impact than others. Vessels under sail obviously create little underwater noise. Low-revving engines have less effect than, say, the high-pitched whine of an outboard motor at speed, which can disrupt cetacean communication and behavior at a considerable distance. Speed itself is a problem: if you are in an area frequented by marine mammals, proceed slowly and keep a sharp

lookout. Also consider having a propeller guard fitted, as whales, dolphins, and manatees are increasingly the victims of propeller strike.

Many people encounter whales while using small power craft, such as the runabouts used by recreational anglers. Operating any craft on open waterways is a serious undertaking, and you should not only be proficient in boat handling, but you should also be provided with suitable safety and communication equipment, and know how to use it. Warm and waterproof clothing, together with sufficient food and drink, are essential.

Once you have sighted whales or dolphins, it's time to abide by the whale-watching regulations (see p. 118). However, there is nothing to stop whales and dolphins being drawn to you. One of the most enjoyable of all whale-watching experiences is to have a vibrant group of dolphins bow riding—jostling for prime position in the pressure wave at the bow; clicking, whistling, and making eye contact with humans peering down from only a yard or so above. If you are moving at high speed

when dolphins approach, slow down. They prefer slower speeds, and will stay with you longer if you do so.

Whales and dolphins seem to enjoy the silent movement of yachts, and dolphins will often bow ride for extended periods with a yacht under sail. Whales are drawn to yachts, perhaps because they are of a comparable size, and are vaguely whale-shaped. If whales are encountered while sailing, you may wish to heave to nearby, and wait to see if they will approach you, as they often will.

COMMERCIAL OPERATORS

Commercial whale watching is a booming industry in many parts of the world, and has been an economic lifeline for many coastal towns, including ex-whaling communities. Most people experience their first close contact with whales and dolphins through such commercial operations, which can provide some truly memorable experiences.

Most commercial operators are responsible, and have a genuine concern for the well-being of the whales. They interpret natural history and behavior for their clients, and most people come away from such experiences with a heightened understanding of and concern for the magnificent animals they have been privileged to see. But because whale watching has become such a competitive industry, ethics are sometimes sacrificed

CLOSE ENOUGH *Kayaker meets orcas (above) in British Columbia, Canada.*

to gain an edge. To provide spectacular displays, such as breaching, for their customers, some operators harass whales. You can exercise responsibility by choosing operators with a good reputation, or by pointing out infractions, should they occur, to the operator.

HYDROPHONES

The calls of cetaceans are an important part of their social interactions. Many people have heard the famous songs of humpback whales, but all cetaceans "vocalize" at various times. The world of underwater sound that can be accessed by hydrophones, underwater microphones, can be a revelation. Apart from the moans, cries, and clicks of cetaceans, there are the beautiful calls of seals, the snapping clicks of shrimp, and the astonishing variety of sounds made by fish. You may also hear how pervasive are the sounds of boats.

The technology for listening in on whales and dolphins is simple and relatively inexpensive. Hydrophones (inset picture) are lowered 20–30 feet (6–9 m) into the water from a stationary boat. These are connected to a pre-amplifier, which magnifies the signal before it passes to a cassette recorder. Complete off-the-shelf units or hydrophone and pre-amp components are readily available from mail-order suppliers of electronic and scientific equipment.

DANGEROUS ENCOUNTERS

The aim of whale watching is to bring cetaceans and humans together,

but there is a risk of physical harm to both during these encounters.

Even dolphins and porpoises, the smallest of whales, are powerful animals compared to ourselves, and can cause humans and their vessels serious injury. There is also great potential for vessels to injure cetaceans.

DAMAGE *A breaching whale (left) can inadvertently wreck a small boat or kill a swimmer.*

BOATS

Violent encounters between whales and boatloads of humans are not a recent phenomenon. For thousands of years in open-boat whaling cultures, whaling was, in effect, combat at close quarters, with loss of life from overturned or smashed boats. Sperm whales and gray whales were thought particularly dangerous by open-boat whalers, but these animals were responding to human attempts to kill them, a far cry from the intent of modern whale watchers.

There are still dangers of physical collision between cetaceans and vessels. These may be completely accidental, or they may be the result of a whale retaliating against what it perceives as a threat from a boat. Many North Atlantic right whales show scars from

ship strike. Boats, especially small, fast runabouts, should never be operated at high speed in an area where whales are likely to be. It is almost impossible for surfacing whales to avoid fast boats that change course rapidly.

In many countries, when you are near whales in any circumstances, you are bound by regulations governing your approach distance and behavior (see p. 118). These regulations simply formalize a code of etiquette based on the natural behavior of cetaceans. Put simply, be

INJURIES *A historical woodcut of a boat striking a whale (above); and a gray whale (left) with propeller scars.*

considerate. Any person who is not, is asking for trouble.

Whales themselves are generally very considerate of small vessels, and have no reason to confront them unless provoked. Provocation need not be intended. A boat innocently bearing down on a submerged whale mother and calf may represent a serious threat to the mother, who, to the surprise of the sailors, may defend herself accordingly. Many tales of "attacks" on yachts and other small vessels may stem from accidental confrontations and collisions, some of which have resulted in sinkings, and probably in dead or badly injured whales.

On the other hand, whales are free to approach, and you may find that although you have kept a respectful distance they are suddenly around you. Whales involved in very active behavior, such as feeding or mating competition, should be given a wide berth. These are important activities that demand the whales' complete attention, and small vessels drifting into the center of such a situation may be in peril. The mere flick of a tail fluke may be enough to destroy a small

presence of swimmers, have awesome power. An experienced Argentine underwater photographer had several bones broken when a southern right whale gave him just a light bump.

Scuba diving with cetaceans is not recommended, because many species don't like the bubbles generated. Some species, such as humpbacks, blow bubble streams as a threat to other whales.

In many countries, intentional swimming with cetaceans is not permitted, but there is nothing to stop cetaceans swimming with you, once you have reached the minimum allowed distance from them. Always remember that there is no truth in the adage that where there are dolphins, there are no sharks—in many areas, sharks feed on dolphins, and even on large whales.

Whales, dolphins, and porpoises are not cute and cuddly; they are intelligent, but very different animals. We have yet to learn what they think of us, if they think of us at all. Meanwhile, we must treat them with the utmost respect, not only because they are inherently such wonderful animals, but also because they have the power to kill or injure us.

craft. In Baja California, Mexico, two people were killed when their whale-watching boat was rammed by a female gray whale. Their boat had strayed between the mother and her calf. Whales will often give warning of their displeasure when boats are too close or acting inappropriately, but you may be unable to read the signals.

Swimmers

In many parts of the world "dolphin swims" have become popular pursuits, but we should be aware that cetaceans don't necessarily share the excitement and pleasure that humans derive from this activity, and may sometimes find it threatening. Although there have been many examples of friendly, even intimate interactions between cetaceans and humans in the water, swimming with dolphins is not without risk. There has been at least one documented case in which a bottlenose dolphin, which had made a habit of approaching swimmers at a Brazilian

beach, fatally rammed a swimmer who is believed to have been harassing it. There have also been cases of captive cetaceans turning on their trainers and killing them.

Many people idealize dolphins almost as toys—benign, smiling creatures that could not possibly wish us harm. Yet in a recent case, a woman swimming with wild short-finned pilot whales was seized by the leg and held underwater almost to the point of drowning. While cetaceans often treat us with curiosity and respect (or indifference), we should never assume that they may not act otherwise for reasons of their own. Large whales, while graceful in the water and extremely aware of the

WHALE-WATCHING GUIDELINES

*Some countries now have whale-protection legislation in place—
with whale-watching guidelines being an important component.*

These guidelines may have the force of law, and are designed to reduce the potential disturbance of cetaceans by vessels and aircraft. One reason such guidelines are necessary is the physical welfare of those aboard the vessels (see p. 114), who may be at risk in situations where whales feel harassed. The main reason, however, is the welfare of the whales, because they are vulnerable to disturbance and harassment, even when it is unintended.

Underwater sound is a very important form of disturbance to cetaceans that is frequently overlooked because it is invisible. Sound travels much faster and farther through water than in air. A cetacean's sense of hearing is acute and it uses sound for communication, navigation, and finding food. Many vessels produce engine or propeller noises that are so loud that they can be detected from miles away, so at close range they could disturb many species. Just as important, some frequencies are more of a problem than others, depending on the hearting and vocal ranges of the species concerned.

It is thought that whales can detect sounds made by an icebreaker 18 miles (about 30 km) distant. Even the sounds of aircraft can penetrate water: a low-flying

DISTRESS SIGNALS *The beluga whale (above left) and the humpback (above) are both blowing underwater, a possible sign of disturbance.*

jet fighter passing overhead was detected by hydrophone at a depth of 60 feet (18 m). With these facts in mind, the potential for disturbing whales by a group of noisy whale-watching vessels or low-flying aircraft becomes more obvious.

PHYSICAL INTRUSION

There is also the issue of physical disturbance. The presence of boats (combined with greatly increased noise levels as the vessel approaches) within or near a group of whales, or even a single whale, can disrupt their normal activities. Whales and dolphins live their lives according to strict time and energy budgets and, often, can ill afford to be distracted from vital behaviors, such as breeding, feeding, or nursing calves, to avoid craft that may be disturbing them.

Boats can physically threaten whales by violating

"polite" distances, or they can separate whales from others in the pod, which is particularly serious in the case of mothers and calves. Low-flying aircraft, particularly helicopters, may appear threatening to cetaceans.

Cetaceans respond to the attentions of humans in different ways. If they are approached carefully and considerately, they may simply carry on as they were, minding their own business. They may decide to break from what they were doing and come to investigate the vessel, which is just what the whale watcher wants. If the attention is unwelcome, they may show it in various ways.

The ability to read these changes in behavior is the key to success, and anyone who

intends to watch whales regularly should acquire the skill. At the first sign of disturbance, such as an unusual or abrupt change in behavior (see below), you should back off immediately.

SIGNS OF HARASSMENT

"Harassment" is persisting in following or approaching whales when they are exhibiting escape or disturbance behaviors. If you are not sure if you are disturbing the animals, give them the benefit of the doubt. Escape or disturbance behaviors include: rapid changes in speed or direction; increased breathing rate; prolonged diving or changing course underwater; blowing underwater, or loud "growling" blows at the shielding of calves by their mothers; tail slapping, tail slashing or breaching; or making threatening "rushes" past the boat. These may all (but not always) indicate that the whales are upset by your presence. One problem is that some of these behaviors are spectacular, and some unscrupulous individuals are known to harass whales deliberately to get a spectacular response. This very exploitative view of cetaceans ignores their needs.

As a member of the whale-watching public, you should regard it as your responsibility to become aware of the relevant standards and behaviors. Try to base your own personal guidelines on the most stringent available.

You may feel moved to request that others, including commercial operators, abide by them. But remember that guidelines are not based on biological realities, they are a compromise that gives the whale-watching public what it wants: a reasonably close view of whales. While the legal minimum approach distance is usually 110 yards (100 m), cetaceans will usually be well aware of you before you reach that point, and may be moving to avoid you. Chasing a whale to bring it to the legal minimum distance may constitute serious harassment. The responsible whale watcher knows when to leave the animals alone.

HOW TO APPROACH A WHALE

Never pursue a whale that seems disturbed by your actions and never separate whales in a pod. If seriously alarmed, whales may change direction, or remain submerged for longer than usual, or even leave the area. Avoid making loud noises. When approaching in a boat, either (A) position your craft 330 yards (300 m) ahead and to one side of the animal and wait quietly for it to pass, or (B) approach slowly from the side, never closer than 110 yards (100 m). Swimmers should leave 33 yards (30 m) clear. (Illustration not to scale.)

A 330 yd

110 yd

B

RESPONSIBLE WHALE WATCHING

Whale watching in itself is not a harmful exercise if conducted responsibly, and with the whales' needs to the fore at all times.

In terms of our relationship to cetaceans, we live in a privileged era. There has never been such freedom to travel the Earth to see them in their natural habitats, and there has never been a greater understanding of these wonderful animals, nor of the problems that beset them.

In the past, commercial exploitation of whales and dolphins has often been appalling. Yet with hindsight, we can see why it was so. Ignorance and greed are powerful forces in human society, and without an appreciation of the intrinsic value of wild marine mammals to counter these forces, exploitation was inevitable.

Over the past 25 years, the situation has changed dramatically, with an explosion in scientific knowledge of cetaceans, and a shift to a much more respectful attitude toward them in many parts of the world. Yet despite the appreciation of cetaceans that we now have, their overall plight worldwide is worsening. The exploitation of many species is escalating and the marine environment on which they depend faces an ever-increasing onslaught.

A MATTER OF ATTITUDE

How does this relate to whale watching, and to our responsibility? It all comes down to our attitude toward whales, dolphins, and porpoises. Are they simply a food source? Some think so. Do we see them as some sort of wild, living theme-park display, to be glimpsed briefly and then forgotten? Or are they intelligent creatures, living lives independent of humans, and having a far older claim to the Earth than we have?

Unfortunately, human activities now intrude on the lives of cetaceans almost everywhere, through whaling, hunting, overfishing, pollution, habitat degradation, and noise in the oceans—to list but a few of the problems. Is whale watching to be added to the list of activities that diminish the quality of cetaceans' lives? Or is it to be something that adds to the value of our own experience of the world, while allowing whales and dolphins the respect and consideration that are their due?

THE BUSINESS OF WHALE WATCHING

Concurrently with specific whale population recoveries in many areas in the past 20 years, whale and dolphin watching has boomed around the world. Cetaceans represent a market niche that entrepreneurs have rushed to fill. In coastal areas along migration routes, and at some feeding and breeding areas,

HARASSMENT *Historically, whales (top) have been treated badly. The watchers (above) are crowding these orcas by surrounding them. The noisy and sudden approach of this boat (left) may be distressing the humpback..*

BETTER ALTERNATIVE *Quietly observing the natural behavior of southern right whales in an organized fashion (above), without disturbing their activities, is surely of greater value than seeing them perform "tricks" at theme parks, as the orca (right) is doing.*

whale watching is now billed as a major tourist attraction, and the number of operators is proliferating. In some areas, the number of licenses is regulated, more often for business reasons or to cap competition, than for the interests of the whales or dolphins.

Many countries now have whale-watching guidelines (see p. 118), to protect cetaceans from overzealous watchers. But when a competitive industry is based on observing wild animals, there will always be tension between commercial pressures and the needs of the animals. All too often, the animals' needs come second.

RESEARCH AND EDUCATION

It is important that cetaceans benefit from whale watching through education, research, and conservation. Educational whale-watch trips make people more sympathetic toward the animals and, ultimately, toward marine conservation.

A common, and highly successful, arrangement operates in North America and some other parts of the world in which biologists act as on-board naturalists. They help to find the whales, provide informative and entertaining commentaries, and answer questions from clients. In return, they have free passage on every trip to do their research and, sometimes, are allowed to sell merchandise on board to raise essential funds. In a few places, the whale-watch operators donate a percentage of the ticket price to buy research equipment or to pay for other aspects of the research work.

THE ROLE OF THE INDIVIDUAL

We all have a responsibility to protect cetaceans from harm or harassment, whether obliged to by law or not. The actions of every boat owner or operator in the vicinity of cetaceans can have an effect, either positive or negative, on their welfare. Careless and ignorant boat handling may not only distract and annoy whales, it can endanger them, or threaten to displace them from an area that they have visited for untold generations. Many boat owners and operators have no idea how to behave around whales, and some shocking acts of harassment can result. We may choose to exercise our responsibilities by objecting to the actions of such operators.

At the least, we can ensure that our own behavior around cetaceans is not likely to have a detrimental effect on their behavior and welfare. We can do this by learning to recognize and interpret the basics of their behavior and social life (which is fascinating in itself), and by familiarizing ourselves with the stringent whale-watching guidelines that exist for this purpose.

As cetaceans do not have a voice to express themselves, we must try to be aware of their requirements, and to minimize the effects we may have on them. Otherwise, we are merely engaging in another form of exploitation, which may contribute to a decline in the ability of these animals to feed, to reproduce, and ultimately, to survive.

CHAPTER SIX
IDENTIFYING WHALES, DOLPHINS, *and* PORPOISES

Greatest of all is the Whale, of the Beasts which live in the waters,

Monster indeed he appears, swimming on top of the waves,

Looking at him one thinks, that there in the sea is a mountain,

Or that an island has formed, here in the midst of the sea.

ABBOT THEOBALDUS (c. 1022)

OBSERVATION TECHNIQUES

Finding whales at sea is quite a challenge, but with practice, patience, and a little luck, anyone can do it.

The first step is to work out exactly where to look. This could be near the coast, over an underwater canyon or seamount, on the edge of the continental shelf, in a region of upwelling or around strong tidal runs, near an estuary, or anywhere that seems appropriate for the species you are hoping to find. Speaking with local fishers, studying charts, and reading about the area can all help you to identify the most likely places.

VANTAGE POINT *Perched aloft on the mast (right), a whale watcher scans the waters of southeast Alaska. Blue whale biologists (below) record sightings and set up their equipment in the Sea of Cortez, Mexico.*

HOW TO LOOK

When it comes to the search itself, you can either move around haphazardly, following hunches, or take a more scientific approach and use transects (pre-planned routes along a grid system). Both methods work with varying degrees of success. Whale scientists prefer transects, especially when they are conducting an official survey, because this system gives directly comparable results from survey to survey. Transects also ensure that no part of an area is missed, although, admittedly, they are not as much fun as zigzagging around where your fancy takes you.

Whichever system you choose, try not to be in too much of a hurry. While it may seem that the more ground you cover the better, in fact, the number of whales you see tends to be inversely proportional to the speed of the vessel. So take your time and search properly.

Stand as high above sea level as possible—in the crow's-nest, on the roof of the wheelhouse, or anywhere that is a good vantage point. Then scan the horizon with binoculars, scrutinize the middle distance, and always check around the boat (it is amazing how easy it is to miss dolphins riding the bow waves, while you are busy gazing far out to sea). Finally, remember to scan 360 degrees in case a whale surfaces behind or to one side. The boat may be moving forward, but that does not mean that everything will be up ahead.

WHAT TO LOOK FOR

People who have spent a lot of time looking for whales and dolphins instinctively recognize the tiniest clues. With large whales, it is normally the blow. This is more visible in some weather conditions than others, but it can be surprisingly distinctive.

To learn is a natural pleasure, not confined to philosophers, but common to all men.

Poetics,
ARISTOTLE (translated by
Thomas Twining)
(384–322 BC), Greek philosopher

It may look like a gradual puff of smoke, or a flash of white.

Alternatively, you may just see the whale's back breaking the surface, or its tail raised briefly before a dive. It often resembles a strange wave that does not look quite right, so anything suspicious is worth taking the time to investigate.

Splashes are also telltale signs. They can be caused by breaching, flipper-slapping, and lobtailing whales, or by a distant school of dolphins, which frequently look like just a rough patch of water.

The presence of birds can be another useful clue, particularly if they are concentrated in a small area, or seem to be feeding. It could mean there are lots of fish in the water, and whales and dolphins may be feeding from underneath.

LISTENING FOR SPERM WHALES

Sperm whales can be difficult to find. Spending most of their lives out of sight in deep water, they behave more like submarines than air-breathing mammals. So, unless you are lucky enough to be in the right place at the right time, the chances of a close encounter are fairly remote.

There is, however, an alternative. With the help of an underwater microphone, or hydrophone (see p. 115) lowered from a boat (as pictured at right), it is possible to

hear them instead. Since sperm whales cannot see where they are going in the dark ocean depths, they build up a "sound picture" of their surroundings by making loud clicking noises and listening for the returning echoes. An experienced operator can track their movements from these clicks, and can even judge where they are likely to surface briefly to breathe. With a little patience, and careful maneuvering of the boat, a close look is almost guaranteed.

THE CLOCK SYSTEM
When you see a whale or dolphin for the first time, there is a temptation to shout "Over there!" or "Look!" But by the time everyone else has seen where you are pointing, and then looked in the right direction, it has usually disappeared. The so-called "clock system" avoids this problem by designating 12 o'clock for the bow, 6 o'clock for the stern, 3 o'clock for starboard, 9 o'clock for port, and so on.

"Whale at 6 o'clock!" for example, means that you have seen a whale immediately behind the boat. After a while, everyone becomes so skilled at this system that they can call an accurate "Whale at 4.36!" and everyone knows exactly where to look.

GETTING CLOSE
In many parts of the world, there are strict regulations governing the way you are allowed to approach whales (see p. 119) and, of course, nothing is more important than their welfare and safety. The golden rule is to always move slowly. When you are within their range, cut the engine and let the whales decide what happens next. Surprisingly often, the whales will surface quite close to the boat—and your reward will be a really close encounter.

ALL EYES *are on this whale, sighted off the coast of Hawaii. If it can be reliably recognized, researchers can gain valuable insights by studying its behavior.*

IDENTIFICATION TECHNIQUES

Identifying cetaceans at sea can be frustratingly difficult, and even the world's experts are unable to identify every animal they encounter.

Developing the skills necessary to tell the species apart can be satisfying, but the animals do not make it easy. They spend most of their lives hidden from view underwater, and even when they come to the surface to breathe, they often disappear before it is possible to get a good look. When they do reveal more of themselves, or stay on the surface for longer, many of them look remarkably similar and are almost impossible to tell apart. There is even a great deal of variation within each species, and no two individuals look exactly alike: they vary in color and size, behave differently, and their

dorsal fins are not uniform.

The simple fact that whales and dolphins live in the sea adds yet more complications. Imagine trying to get an accurate impression of an animal on the move, while struggling to keep your balance on a rolling, slippery deck. Adverse conditions, such as a whitecaps, high winds, heavy swell, driving rain, or even glare from the sun, add to the challenge.

Despite all these potential difficulties though, there are ways of identifying whales, dolphins, and porpoises. In fact, it is quite possible for anyone to recognize the relatively common and distinctive species and,

eventually, many of the more unusual ones as well. All it requires is a little background knowledge and some practice.

The best approach is to use a relatively simple process of elimination. Run through a mental checklist of key features every time a new animal is encountered at sea. The more features you are able to take into account, the better chance you have of making a positive identification. Ultimately, this process becomes almost automatic, and you learn to recognize a particular species by its unique combination of features or, in bird-watching jargon, its own "jizz," a slang term for "general impression of size and shape."

CHARACTERISTICS *of the pygmy right whale (below, top), typical of baleen whales, could never be confused with those of toothed whales, such as the Atlantic white-sided dolphin (below, bottom).*

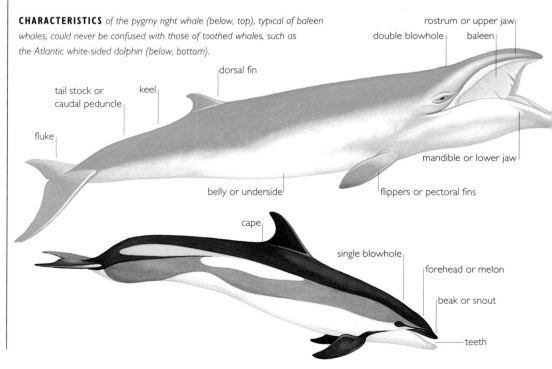

- dorsal fin
- tail stock or caudal peduncle
- keel
- fluke
- belly or underside
- rostrum or upper jaw
- double blowhole
- baleen
- mandible or lower jaw
- flippers or pectoral fins
- cape
- single blowhole
- forehead or melon
- beak or snout
- teeth

NOTEWORTHY *The tall dorsal fins of three males (right) identifies this as a pod of orcas. The distinctive heart-shaped or V-shaped blow (far left) is characteristic of the gray whale. The prominence of the beak of this common dolphin (below) is one of its distinguishing features.*

IDENTIFICATION CHECKLIST

Becoming familiar with the most obvious features of your subject will help greatly.

Geographical location There are not many places in the world with records of more than a few dozen species of whales, dolphins, and porpoises. So taking the location into account immediately limits the number of possibilities.

Habitat Just as woodpeckers inhabit woodlands rather than shores, and giant pandas prefer mountains to wetlands, most whales, dolphins, and porpoises are adapted to specific marine or freshwater habitats.

Unusual features The tall dorsal fin of the male orca, the long tusk of the male narwhal, the extraordinary curved teeth of the male strap-toothed whale, and highly distinctive features of some other species can often be useful for a quick identification.

Size Estimating size accurately at sea is notoriously difficult, unless a direct comparison can be made with the length of the boat or something else in the water. But simply deciding whether the animal is small, medium, or large can help to eliminate a range of possibilities.

Color and markings Distinctive markings and bright colors can be useful identification features. However, since colors at sea vary according to water clarity and light conditions, they can also be quite confusing if they are at all subtle.

Dorsal fin The size, shape, and position of the dorsal fin varies greatly among species and a few species have no fin at all. While rarely enough alone, such details can be useful in combination with other features for making a positive identification.

Flippers It is rarely possible to see the flippers, or pectoral fins, clearly—but their length, color, shape, and their position on the animal's body, vary greatly from one species to another.

Body shape The animal's overall shape can sometimes be useful, although many whales, dolphins, and porpoises rarely show enough of themselves to give a satisfactory impression.

Beak In oceanic dolphins, and some other toothed whales, the presence or absence of a prominent beak is a particularly useful identification feature.

Flukes Some large whales lift the flukes high into the air before a dive, others do not. This distinction alone can help with identification, but it is also worth checking the shape of the flukes, which vary considerably from species to species.

Blow or spout The blow is immensely useful for identifying larger whales. It varies in height, shape, and visibility among species and, especially on calm days, can be all that is needed for a positive identification.

Dive sequence The way in which a whale, dolphin, or porpoise breaks the surface to breathe, and then dives again, is known as its dive sequence. In some species, this is very distinctive, although differences in others can be quite subtle.

Behavior Most whales, dolphins, and porpoises have been known to breach, pec-slap, lobtail, and perform other well-known behaviors at one time or another. But they all have slightly different techniques, and some are more active at the surface than others.

Group size The number of animals seen together is a useful indicator; some species are highly gregarious while others tend to live alone or in small groups.

DEVELOPING
IDENTIFICATION SKILLS

Whale biologists and expert naturalists on whale-watch tours somehow seem able to identify every fleeting blow, fin, back, or tail they see—merely at a glance. So what is their secret?

Encountering a whale, dolphin, or porpoise at sea, and then not being able to make a positive identification, can be incredibly frustrating. Not surprisingly, there is often a temptation to guess. Some people play it safe, and opt for the most likely species for the region; others tend to be more optimistic and go for one they have not seen before.

But this is a bad mistake. The first rule of whale identification is: never jump to conclusions. Professional whale scientists don't, because it makes their survey and research results unreliable, and whale watchers shouldn't, because it does little to improve their identification skills. It is perfectly acceptable to record simply "unidentified dolphin" or "unidentified whale," if a more accurate identification is not possible. In fact, if you check the log of any official whale survey, the chances are there will be at least a few unconfirmed sightings recorded in this way.

RECORDS *On-the-spot notes jotted in a notebook (above) can be checked later from reference books. Photos of behavior, such as a blue whale fluking (right), are invaluable.*

KEEPING NOTES

Having logged your sighting as "unidentified," it is important to write notes. Learning to make detailed, accurate field notes is another basic essential of whale identification (see p. 108). As well as helping to improve your field skills, it may help you to make a positive identification if you see the same species again, whether it be hours, days, weeks, months, or even years later. Better still, it may enable you to identify the animal at home, where you can take your time and consult field guides and your other reference works.

The sooner you write the notes, the better. It is amazing how quickly many of the more subtle details of a sighting are forgotten, even just a few minutes after a whale or dolphin has

disappeared into the murky depths. Run through the list of identification features described briefly on page 127, from the dive sequence to the shape of the beak, and then make a note of the time, date, location, weather, and sea conditions, and any other details that might be relevant.

At first, it is better to err on the side of making too many notes, even if you seem to be spending more of your time writing about the whales than watching them. After a little practice, of course, it will become easier to decide what you should or should not include. But never stop writing and sketching altogether. Even as you become more proficient at whale identification, it is still worth jotting down useful

A stander-by may ...,

perhaps, see more of the

game than he that plays it.

A Tritical Essay Upon the Faculties of the Mind, JONATHAN SWIFT (1667–1745), English satirist

COLLECTING DATA *Observing belugas in the Gulf of St. Lawrence (top, far right). The unmistakable fluke of a humpback (above). A naturalist (right) points out whale features to look for during a tour in Samaná Bay, Dominican Republic.*

details. They all help to build up a more complete picture of the species you see regularly. You could include photographs and other mementos to turn your journal into a permanent record of your sightings, a valuable long-term reference source and an attractive journal to treasure in years to come.

LEARNING FROM THE EXPERTS

Reading books, watching films about whales, studying photographs, and even sharpening your powers of observation on local wildlife, will all help to hone your identification skills. But there is no real substitute for whale watching with an expert. This is one reason why it is so important to select whale-watching tours with knowledgeable biologists or naturalists on board.

Listen to the commentaries and don't be shy about asking

questions—most people working on whale-watch boats enjoy talking about whales and welcome questions from anyone with a genuine interest.

After a while, it will become clear that people who spend a substantial amount of time around cetaceans begin to develop a feel for the "jizz," or general impression, of different species. Subconsciously, we all use a similar technique for recognizing one another. We recognize our friends and relatives instinctively because of an indefinable combination of

some or all of their features. With wildlife, using the jizz is not always a foolproof system—and is no substitute for checking the finer details later—but it is definitely a technique worth taking the trouble to master.

Becoming proficient at whale identification is much like developing any other skill. It takes time, perseverance, and practice. At the end of the day, it is all about attitude of mind: watching inquisitively, listening carefully, writing notes properly, and identifying cautiously.

BODY SHAPE *and* SIZE

Whales, dolphins, and porpoises have broadly similar body shapes, but they vary greatly in size and, on closer inspection, reveal many subtle differences.

IDENTIFYING FEATURES
A clear view of the flipper of a short-finned pilot whale (above). A bottlenose whale (left) has a bulbous fore-head and dolphin-like beak.

On land, it is possible to estimate the size of animals with considerable accuracy, using trees, telegraph poles, bushes, rocks, buildings, and a host of other recognizable objects to provide the essential element of scale. But it is notoriously difficult to estimate the size of a whale or dolphin, a bird, a boat, or anything else in a vast expanse of "empty" sea. If 10 people were asked to estimate the size of a whale or dolphin, chances are there would be 10 completely different answers.

If the animal comes close enough to the boat, it may be possible to make a direct comparison: a whale roughly two-thirds the length of a 100 foot (30 m) vessel, for example, is clearly about 65 feet (20 m) long. But most of the time, it is necessary to estimate the size without the benefit of such a comparison.

Despite the difficulties involved, this is not a major problem for identification

purposes. The trick is simply to decide whether the animal is small (up to 10 feet [3 m]), medium (10–33 feet [3–10 m]) or large (more than 33 feet [10 m]). Don't try to be accurate to the nearest few inches, or even feet, because this scale will instantly narrow down the number of possibilities: 42 species are "small," 28 species are "medium," and only 11 species are "large."

BODY SHAPE
Unfortunately, many whales, dolphins, and porpoises rarely show enough of themselves to reveal their overall body shape. But when they leap into the air, or are visible underwater, this can be an immensely useful aid to

NEW ZEALAND'S *Hector's dolphins (left) have unusually rounded dorsal fins.*

identification. The huge, rotund body of a right whale, for example, is quite different to the long and relatively slender body of a fin whale. Many of the differences between species are far more subtle to the untrained eye, but with a little practice, they can easily be recognized.

DORSAL FINS
The character of the dorsal fin is rarely enough on its own for a positive identification, especially as it varies greatly from one individual to the next, but it can be useful in combination with other features. The key variables to look for are its size, shape, and position on the animal's body.

The blue whale's dorsal fin is distinctive for being small and stubby, and for its location three-quarters of the way along the animal's back. In

Leviathan …

Upon earth there is not

his like …

The Bible,
BOOK OF JOB 41:33

contrast, the sei whale has a tall, sickle-shaped fin, located much farther forward. The male orca has an unmistakably tall dorsal fin, reaching an incredible height of up to 6 feet (1.8 m), while the dorsal fin of many hump-backed dolphins sits on an elongated hump in the center of their back. Hector's dolphins have distinctive rounded fins that are likened to Mickey Mouse ears.

Only eight species have no dorsal fin at all: the finless porpoise, the southern right whale dolphin, northern right whale dolphin, narwhal, beluga, northern and southern right whales, and the bowhead whale. A number of others,

including most of the river dolphins, the Irrawaddy dolphin, sperm whale, and gray whale, have only tiny dorsal fins or small humps.

FLIPPERS (PECTORAL FINS)
It is rarely possible to get a clear view of the flippers, or pectoral fins, when watching whales, dolphins, and porpoises at sea. Humpbacks sometimes wave their enormous flippers in the air, which is handy for identification purposes, but they are the exception rather than the rule. In most species, the flippers remain well hidden—except during a breach or a close encounter when the water visibility is particularly good.

However, the length, color, and shape of the flippers, as well as the position on the animal's body, vary greatly from one species to another and can be useful in conjunction with other identification features. They

can be short or long, narrow or broad, drab or strikingly colored, and come in a range of combinations in between.

Interestingly, the length of the flippers is one of the few ways of distinguishing a short-finned pilot whale from a long-finned pilot whale at sea.

BEAK OR NO BEAK
The presence or absence of a long, well-defined beak is a valuable identification feature in many toothed whales. In some species, the beak is distinctive and clearly demarcated from the forehead, while in others it is barely noticeable or even absent. Pilot whales, for example, have bulbous foreheads with indistinct beaks, while bottlenose whales have similar foreheads with dolphin-like beaks.

The 27 species of oceanic dolphin are difficult to tell apart at sea, but simply checking for the presence or absence of a beak immediately cuts the possibilities by half.

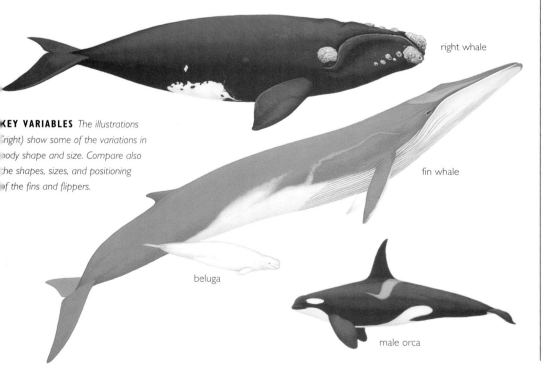

right whale

fin whale

beluga

male orca

KEY VARIABLES *The illustrations (right) show some of the variations in body shape and size. Compare also the shapes, sizes, and positioning of the fins and flippers.*

DIVE SEQUENCE

*The way in which some whales, dolphins, and porpoises break
the surface of the water to breathe, and then dive again,
can be sufficiently distinctive to make a positive identification.*

The dive sequence is particularly useful when identifying large whales, although there are many factors to take into account. For example, the way a whale breaks the surface can vary according to its age, if it is feeding or engrossed in breeding activities, the speed at which it is traveling, the depth from which it has surfaced, and whether it is feeling stressed or relaxed.

The sequence also varies if the whale is simply catching its breath or about to embark on a deeper dive. A gray whale, for example, may break the surface to breathe half a dozen times before diving, but may raise its flukes only on

the final occasion. Normally, it is this final dive sequence that is the most revealing for identification purposes.

Among the large whales, seven species regularly lift their flukes high before diving: gray, blue, bowhead, northern and southern right, humpback, and sperm. Raising the flukes is a useful identification feature in itself although, unfortunately, none of these whales fluke every time; on some dives they prefer to keep their flukes below the surface. The shape and design of the flukes can also be useful for identification, since they vary considerably from species to species. Sperm whales, for example, have broad, triangular flukes, while

gray whales have flukes which have noticeably convex trailing edges.

Bryde's, minke, sei, and fin whales almost never lift their flukes on a dive. Inevitably, though, there are exceptions: several fin whales in the Sea of Cortez, Mexico, for example, lift their flukes on a regular basis.

MINKE WHALE
The minke whale has a very distinctive dive sequence. First to appear is normally its sharply pointed snout, which breaks the surface at a slight angle. As the rest of the head comes into view, it drops to a much shallower angle, the whale begins to blow and

DIVE SEQUENCES
*The dive sequence is often a useful tool
for identification purposes.*

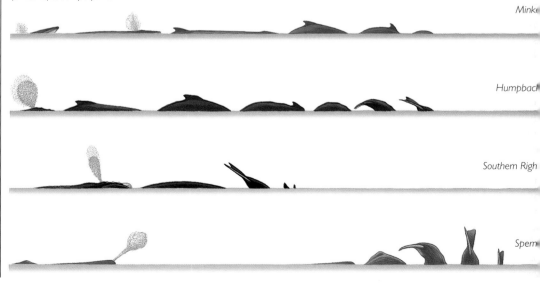

Minke

Humpback

Southern Right

Sperm

then the blowholes appear. The blowholes and dorsal fin are often visible at the same time, distinguishing the minke from all other members of its family except the sei whale and smaller youngsters of other species. Toward the end of the dive sequence, the back and tail stock begin to arch strongly, in preparation for a long dive, but the flukes do not appear above the surface.

HUMPBACK WHALE

When a humpback whale surfaces to breathe, its splash guard and blowholes appear above the surface first. Then its low, stubby dorsal fin comes into view, and its distinctive sloping back forms a shallow triangle against the surface of the sea. As the humpback arches its body, forming a much higher triangle, the hump on its back is especially evident. Toward the end of the dive sequence, while the dorsal fin drops below the surface, the tail stock is strongly arched and the whale rolls forward. As it steepens its angle of descent, its flukes begin to appear above the surface and the distinctive black-and-white markings on the underside are clearly visible from behind.

SOUTHERN RIGHT WHALE

The head of a southern right whale is usually held high out of the water as it surfaces to breathe, showing its distinctive callosities. After the whale has produced its characteristic V-shaped blow, its head disappears below the surface and only the smooth, broad, finless back is visible. Free of barnacles and callosities, the back is also very distinctive. The whale normally raises its flukes when diving, but sometimes may "false-fluke" and not lift them out of the water.

SPERM WHALE

The sperm whale frequently remains motionless when catching its breath, although it may swim leisurely. As it lifts its head for a final breath, only two-thirds of its body length is normally visible. The sperm whale then straightens and stretches its body, gently

arches its back and briefly disappears. It accelerates and reappears a little farther on. Arching its back until it is high out of the water, making the rounded hump and "knuckles" along the upper side clearly visible, it then throws its flukes and the rear third of its body high into the air and drops vertically, with barely a ripple.

SMALL CETACEANS

Some small cetaceans also have distinctive dive sequences. The harbor porpoise often rises to breathe in a slow, forward-rolling motion, for example, making it appear as if its dorsal fin is mounted on a revolving wheel; both the Indus and Ganges river dolphins may surface at such a steep angle that their entire head and beak can be seen above the surface; and the dwarf sperm whale simply drops out of sight, rather than rolling forward at the surface like most other small whales.

While it can be immensely useful, the dive sequence is not a foolproof means of identification. The trick, as always, is to use it in combination with other features.

BLOW SHAPE *and* SIZE

Experienced whale watchers can often tell one species from another just by the height, shape, and visibility of their blows, or spouts.

"Thar she blows!" was once a familiar cry from the mastheads of whaling vessels. For a long time, whalers believed that the animals squirted water from their heads. Pliny the Elder went so far as to suggest that a sperm whale could fill a boat with water from its spout to capsize it. People were often warned to keep out of the spray, which was claimed to be so unbelievably smelly it could cause disorders of the brain. According to another popular myth, if touched by the spray, human skin would peel away as if burned by a flame. Even Herman Melville commented that "if the jet is fairly spouted into your eyes, it will blind you."

WHAT IS THE BLOW?

Despite the myths, it is not water being spouted from a whale's head but air. As the animal surfaces to breathe, its blowhole opens and there is an explosive exhalation followed immediately by an inhalation. The blowhole closes tight once again, and the whale dives. The process takes no more than a few seconds, even in the largest of whales.

BLOWHOLES *Toothed whales, such as the bottlenose dolphin (left), have a single blowhole; baleen whales, such as the humpback (below), have two.*

No one really knows what makes the blow so visible. It is normally whiter and easier to see in cold weather, which suggests that it probably includes water vapor that condenses in cold air. It also contains a small amount of sea water, originally trapped in the area around the blowhole, which is then blown into a fine spray. A third ingredient may be oil droplets and mucus from the whale's air sinuses, windpipe, and lungs—producing a kind of giant sneeze.

IDENTIFYING THE BLOW

Small whales, dolphins, and porpoises tend to have small, brief blows that are rarely visible and normally do not have a recognizable shape.

In contrast, most large whales have very distinctive blows and, in calm weather, these can be recognized even from a considerable distance.

There are a number of variables, however, that must be taken into account. In wet or windy conditions, the shape of the blow can easily be distorted. Even in good weather, V-shaped blows may look like single columns if viewed from the side. Also, the power, height, and shape of a blow varies according to the size of the whale, what it is doing, how long it has been underwater, and so on. The first exhalation after a deep dive may be a thunderous blast heard nearly a mile away, while subsequent blows are less powerful and may even differ slightly in shape.

MARVELS *A nineteenth-century engraving (left) of a whale blow from a German picture book; and the side-on view of the blow (far top left) of the gray whale.*

Distinctive whale blows (head-on view)

BRYDE'S WHALE

The thin, hazy blow produced by Bryde's whale is variable in height, although, typically, it rises to about 10–13 feet (3–4 m); it is rarely distinctive from a distance, especially if the whale begins to exhale before breaking the surface.

SEI WHALE

Sometimes described as shaped like an inverted cone, or a pear, the sei whale's blow appears as a single, narrow cloud; it resembles the blows of fin and blue whales, but reaches a height of only about 10 feet (3 m). It is also not as dense.

SPERM WHALE

The sperm whale's unique, angled blow is extremely distinctive, especially on windless days, and projects to a height of 6½ feet (2 m) or more. Since the blowhole is on the extreme left-hand side of the forehead, the blow is projected forward and to its left.

BOWHEAD WHALE

With two widely separated blowholes, the bowhead whale produces a rather bushy, V-shaped blow; the two diverging clouds of spray rise directly upward to a height of about 23 feet (7 m).

FIN WHALE

The fin whale's strong, straight blow appears as a tall, narrow column of spray; it is usually 13–20 feet (4–6 m) high, and can be seen from a considerable distance.

RIGHT WHALES

The right whale produces a distinctive V-shaped blow, having widely separated blowholes like the bowhead. The plumes rise to a maximum height of about 16½ feet (5 m).

BLUE WHALE

Blowing as soon as its head begins to break the surface, the blue whale produces a spectacular column of spray rising to 29½ feet (9 m) or more. The tallest and strongest blow of all whales, it is noticeably slender and upright.

GRAY WHALE

The gray whale's bushy blow is normally described as V-shaped, but sometimes the two plumes meet in the middle and produce a distinctive heart shape; they normally rise to a height of 10–15 feet (3–4.5 m).

HUMPBACK WHALE

Rising to a height of about 8–10 feet (2.5–3 m), the single bushy blow of the humpback whale is surprisingly visible and distinctive; it is usually wide relative to its height, and can occasionally be clearly V-shaped from a distance.

COLOR *and* MARKINGS

It is sometimes possible to identify cetaceans by their background colors, body patterns, and markings such as stripes or eye patches.

Some whales, dolphins, and porpoises have such unique colors that they are virtually unmistakable. The ghost-like, uniformly light color of the beluga is a particularly good example: superbly camouflaged for life in the Arctic, this whale can be surprisingly difficult to pick out among whitecaps or floating ice, but its white, grayish white, or yellowish white body makes it extremely easy to identify.

Other species can be recognized by a distinctive marking or unique combination of markings. There is nothing quite like Heaviside's dolphin, for example. The front half of its body is uniformly gray and the rear half predominantly blue-black. It has dark flippers, dark patches around its eyes and blowhole, white "armpits," and a white underside; there is even an unusual white finger-shaped lobe on either side of its tail stock. Not surprisingly, the combination of these features alone is usually enough for a positive identification.

The patterns formed by distinctive markings are sometimes as significant as the colors themselves. There are two recognized species of common dolphin, for example, and many color variations, but they all share one feature: an elaborate crisscross or hourglass pattern on either side of the body.

EFFECTIVE CAMOUFLAGE *While difficult to spot among ice floes, belugas (above) are easy to identify because of their distinctive uniformly light color.*

There is a single point where the dark color under the dorsal fin, the tan or yellowish side patch, the pale gray tail stock, and the white or creamy underside all meet in the middle—and it makes a splendidly reliable identification feature. The only other species with an hourglass pattern on its sides is the appropriately named hourglass dolphin, but its hourglass is rather crude in comparison, and is in black and white.

Some markings, of course, are a little more tricky to use for identification purposes, either because they are

especially subtle or because they are not always visible. Minke whales, for example, are broadly similar in color to several of their larger relatives, but many (particularly in the Northern Hemisphere) have distinctive white bands on their flippers. The size and shape of the bands vary, and they are often hidden from view, but in a good sighting, they can be enough to make a positive identification.

SIGHTING CONDITIONS
Unfortunately, colors at sea vary according to the water clarity and light conditions.

IDENTIFICATION
*The minke whale
(left) has a distinc-
tive white band
on the flipper,
especially in the
Northern Hemi-
sphere. The extent
of black and white
coloring (below)
varies among
Dall's porpoises.*

Even in fine weather, these can change dramatically from hour to hour, day to day, and season to season. The same whale may appear quite different, depending on whether it is observed at dawn, in the bright midday sun, or after sunset. Alternatively, if the sun is low on the horizon, it may cast dark shadows over the whale, giving an impression of stripes or other markings. Similarly, if the whale is backlit, it will appear much darker than if the sun is shining from the front. Cloudy conditions, or rain, may alter its appearance yet again.

One of the best examples of this variability is the background color of the blue whale. It is a pale blue–gray color but, at the surface, it often takes on the color of the sea and sky. It can probably reflect more colors than any other whale: as the sun drops below the horizon, and darkness falls, a blue whale may turn from blue to yellow and orange, then red, lavender, dark gray and, ultimately, to black.

INDIVIDUAL VARIATIONS
Another potential pitfall is that members of the same species do not always look

exactly alike (see p. 158). Although they do not, as many birds do, change their markings in the breeding season, their markings may alter as they grow older. Males and females sometimes have different colors and patterns on their bodies, and their appearance may also vary on an individual basis. To identify beaked whales by color alone, for example, is risky, because there is so much variation from one animal to another (and besides, little is known about the coloration of live beaked whales).

As always, caution is essential, even with species that appear to have distinctive background colors, body patterns, and markings. Dall's porpoises, for example, are

very striking: they are predominantly jet black, with a prominent white patch on their belly and sides. But this white patch varies enormously in size—resulting in a wide range of individuals from all black to all white, and with an endless number of variations in between. So just because an animal does not look exactly like an illustration in a field guide, do not assume that it is another species altogether. Just to be sure, check other diagnostic features as well.

DISTINCTIVE CRISSCROSS *All common dolphins can be identified by the elaborate hourglass pattern on their sides.*

UNIQUE FEATURES

Some whales, dolphins, and porpoises are instantly recognizable

by unique features such as scarring, unusual teeth, or strange growths.

It is always wise to check more than one feature before making a positive identification. But a number of species are so distinctive that, with a reasonable sighting, they cannot easily be confused with anything else. Following are some of the most striking examples.

STRAP-TOOTHED WHALE

Female strap-toothed whales are almost impossible to identify at sea, but the males are among the few beaked whales that are really distinctive. They have two extraordinary teeth (see p. 163), which grow from the middle of the lower jaw, curl upward and backward, and then extend to wrap around the top of the upper jaw.

In older animals, these teeth can grow to a length of 12 inches (30 cm) or more, and may actually meet in the middle. When this happens, they form a muzzle, preventing the whales from

STRIKING FEATURES
The vertical crease in the forehead of a Risso's dolphin (left) is unique. The huge, squarish head of a sperm whale (below).

opening their jaws more than a couple of inches.

Despite this apparent handicap, males can still catch squid, their main prey, by using their mouths like vacuum cleaners. Indeed, the teeth may even help—by acting as "guardrails" to keep the food on a direct path to their throats.

RISSO'S DOLPHIN

Any good-sized dolphin that looks distinctly battered is likely to be a Risso's dolphin (see p. 183). The extensive body scarring on adults is caused by the teeth of other Risso's dolphins and, to a lesser extent, by confrontations with squid. Young animals are relatively unmarked and the scarring becomes more pronounced with age.

Risso's dolphin also has a deep vertical crease down the center of its forehead, which is visible at close range and is another feature unique to this species.

WEAPON *The distinctive, twisted left tooth of the narwhal (left) is commonly broken off as a result of fighting.*

SPERM WHALE

Even though it rarely shows much of itself at the surface, the sperm whale (see p. 156) is easy to identify at sea. It is the only large whale with an enormous, squarish head (typically measuring about a third the length of the body). This is particularly huge and distinctive in adult males.

It has three other unique features: it is the only large whale with a single blowhole, an angled blow, and an unmistakable prune-like skin.

NARWHAL

The male narwhal has two teeth (see p. 190). The one on the right normally remains invisible, but the one on the left grows to a remarkable length. It pierces the animal's upper lip, develops into a long tusk and eventually looks rather like a gnarled and

DISTINCTIVE APPENDAGES *A good view of the humpback's unmistakably long flippers (above) as it breaches.*

twisted walking stick. (When viewed from the root, it always spirals in a counter-clockwise direction.) It is probably used in establishing sexual dominance, rather like the antlers in male deer.

At least a third of all narwhal tusks are broken, but unbroken tusks reach an average length of about 6½ feet (2 m) and, in extreme cases, may exceed 10 feet (3 m). Approximately 1 male in 500 has two tusks and 1 female in 30 grows a single tusk.

NORTHERN AND SOUTHERN RIGHT WHALES

The two right whales (see pp. 147 and 148) are instantly recognizable by the areas of roughened skin on their enormous heads. These are called callosities, and occur on the rostrum, the chin, the "cheeks," and the lower lips, as well as above the eyes and near the blowholes. Their function is unclear, although

they may be used in some way in aggressive interactions.

Realistically, the only way to tell a northern right whale from its southern relative is by the geographical location; fortunately, for identification purposes, there is no overlap in their range.

HUMPBACK WHALE

The humpback whale's exceptionally long flippers, or pectoral fins, are unmistakable. They typically measure about 15 feet (4.6 m) in large animals, but grow up to one-third of the length of the whale—making a potential

From space, the planet is blue, / From space, the planet is the territory, / Not of humans, but of the whale.

Whale Nation,
HEATHCOTE WILLIAMS
(1941–), English poet

maximum of more than 18 feet (5.5 m). No other whale has flippers anywhere near as long (see p. 155).

Humpbacks also have unique black-and-white markings on the underside of their tails, which are clearly visible when they raise their flukes before a deep dive.

BRYDE'S WHALE

Bryde's whales and sei whales are strikingly similar in both size and appearance. At a distance, they can be extremely difficult to tell apart and, for many years, were frequently confused. However, Bryde's whales are unique in having three longitudinal ridges on their head (other baleen whales in the same family have just one) and are therefore unmistakable at close range (see p. 151).

The prominence of the two outer ridges is, unfortunately, variable and, in some individuals, one or both may be difficult to detect at sea. So it is always wise to look for other identifying features as well—just to be certain.

IDENTIFYING INDIVIDUAL CETACEANS

Once you can identify the different species, the next step is to recognize particular individuals by variations in their natural markings.

Biologists have been identifying individual animals as part of their research for many years. By observing the same individuals day after day, and year after year, they find it is possible to: follow their movements, map out their home ranges, learn about their special partnerships, work out the sequence and timing of key events in their lives, calculate average life spans, and study anything from group stability to individual idiosyncrasies.

Jane Goodall, for example, used facial patterns to recognize all the chimpanzees in her classic study in Gombe National Park, Tanzania; she found that each member of the troop had a distinctive "haircut," unique eyes and ears, and a recognizable nose and mouth. More recent studies have used the stripe patterns in zebras, the arrangement of whiskers in lions, and even the shapes, nicks, and scars of elephants' ears, to tell one animal from another.

IDENTIFYING INDIVIDUAL WHALES

Whale scientists use similar techniques, identifying individual whales, dolphins, and porpoises in many different ways, depending on the species and the nature of the project. Blue whales, for example, have an individual pattern of mottling on the body and the shape of the dorsal fin varies; right whales can be recognized by the pattern of callosities on the

head; orcas are identified by a combination of the dorsal fin and gray saddle behind; and Risso's dolphins have distinctive scar patterns on their bodies, apparently acquired during fighting.

Humpback whales are recognized by the unique black-and-white markings on the underside of their tails, which range from jet-black to pure white and include an endless number of variations in between. Since no two humpbacks have identical tails, these markings are the equivalent of our own fingerprints. With experience, it is possible to tell one individual from another simply by peering underneath as they lift their tails high into the air in preparation for a deep dive.

In fact, many field biologists have become so proficient at recognizing individuals that "their"

"MUG SHOTS" *Individual Risso's dolphins (above) show unique patterns of body scarring. Patterns of callosities on the head of a southern right whale (left) will aid the identification process.*

animals might as well be carrying passports. In some cases, they even know them all by name. A good example is New England in the United States, where most of the humpbacks have names that describe the patterns on their tails, attributed to jog the memories of the people studying them. Cat's Paw, for example, has a jet-black tail interrupted only by a white paw print on the left-hand side; Seal, as one might expect, has an image of its namesake on the underside; and Fracture has a black tail, with a distinctive white line down the middle.

NO MISTAKE *On blue whales, the pattern of mottling and the shape of the dorsal fin distinguish between individuals.*

PHOTO-IDENTIFICATION

In some whales, dolphins, and porpoises, the differences among individuals can be quite subtle, and some animals are difficult to approach closely, or are so small and fast that it is hard to see details of any markings. At best, it takes a trained eye to tell them apart; at worst, it is an almost impossible task.

Since the risk of human error is so high, most whale biologists do not rely merely on sight and memory. They prefer to take photographs of each animal instead. These so-called "mug-shots" help to confirm their identity, as well as to provide a permanent record of their markings. They are particularly useful when researchers are discussing individual animals.

This invaluable research technique, known as "photo-identification," or simply "photo-ID," has dramatically extended our knowledge of the behavior and habits of wild whales, dolphins, and porpoises in recent years.

SALT: THE MOST STUDIED WHALE IN THE WORLD

A humpback called Salt has been studied by hundreds of whale biologists and encountered by many thousands of whale watchers. Her name was derived from a distinctive white patch and a sprinkling of white spots on her dorsal fin (in the days before anyone had thought of using the black-and-white markings underneath the tail for this purpose). She was among the first of thousands of humpback whales to be identified officially. She has appeared in the Gulf of Maine in the United States every summer since she was first spotted on 1 May 1976.

For several years, no one knew whether she was male or female. It was not until 1980, when she turned up with a calf by her side (later named Crystal for being a little piece of Salt) that she disclosed her well-kept secret. Salt has had many calves since, including Thalassa, Halos, Bittern, and Salsa; and, in 1992, when Thalassa had a calf of her own, Salt became a grandmother for the first time.

In 1978, Salt was photographed off the coast of the Dominican Republic and entered whale lore, by confirming the link between the humpback feeding grounds in the Gulf of Maine and their breeding grounds 1,500 miles (2,400 km) away in the Caribbean. Biologists had long suspected such a link, but Salt provided them with the first firm evidence, a major breakthrough in humpback whale research.

CHAPTER SEVEN
WHALES, DOLPHINS,
and PORPOISES
FIELD GUIDE

Of all animals, none holds sway over so vast a dominion;

their watery realm extends from the surface

to the bottommost depths of the sea.

Histoire naturelle des cétacés,
BERNARD GERMAIN LACÉPÈDE (1756–1825), French naturalist and author

USING *the* GUIDE *to* CETACEAN SPECIES

Getting to know whales, dolphins, and porpoises in all their extraordinary shapes, sizes, colors, and perfectly adapted behaviors is a pursuit that can easily become a rewarding obsession.

The field guide describes and illustrates some of the many species of whale, dolphin, and porpoise that dwell in the world's oceans, estuaries, and rivers. It describes important features and behaviors that might help you to identify an animal you have sighted or photographed, and gives the range in which you might expect to find each species. The more information you have, the better you will be able to appreciate the cetaceans you encounter.

*The **main photograph** shows as much of the species as possible, given the constraints of photographing such wild creatures in circumstances and conditions that can be difficult.*

*The **common** and **scientific names** of each species. The species are grouped according to taxonomy.*

*The **text** provides information on the appearance of the whale, dolphin, or porpoise; its common names; habitat; and where and what to look for when you want to find it. The text gives ideas on how to differentiate between similar species, and supplies information about the cetacean's migratory and feeding habits, and breeding grounds. Also discussed are the history of its encounters with humans, such as whalers, and the extent to which these encounters and environmental damage have depleted the cetacean populations.*

Delphinidae: Oceanic Dolphins

Short-finned Pilot Whale
Globicephala macrorhynchus

FIELD NOTES

■ Male: 20–22' (6–6.7 m); female: 17–18' (5.1–5.5 m); at birth: 4½'+ (1.4 m+)
■ Squid
■ Offshore
■ Warm temperate and tropical world waters
■ Insufficiently known, exploited to some extent, but not extensively

Short-finned pilot whales range through offshore, warm temperate and tropical seas, making deep dives mainly to feed on squid. Pilot whales are so intensely social that they are almost never seen alone.

Identifying short-finned pilot whales is fairly easy. They sometimes travel with dolphins, such as bottlenose dolphins, but they are much larger and their black bodies are darker—most dolphins tend to be gray. The short-finned pilot whale's dorsal fin is broad based and set far forward on the back. Its head is rounded and bulbous. The only animal likely to be confused with this whale is its close relative, the long-finned pilot whale (see opposite page).

Short-finned pilot whales resident in the warm waters off Tenerife in the Canary Islands have been studied, with 445 individuals photo-identified as part of

long-term behavioral rese Here, scientists have foun the whales prefer water of about 3,300 feet (1,0 Typical group size is 10 but some pods are as la Like orcas, pilot whale long time, with female to survive up to 65 ye Short-finned pilot rarely breach, but so spyhop. They can be whale-watch trips in the Bahamas; eastern Caribbean; in the Azores; i Prefecture, off Japan; in the Tañon the Philippines; and around Hawa Although not hunted to the sa long-finned pilot hundred short-fi whales are kille Japan, as well a eastern Caribb fishers. In Japa are two forms Both pilot some variati of populatic may lead to another sp

Short-finned pilot whales with a yearling calf, Azores Islands.

188

*The **common** and **scientific names** of each species. The species are grouped according to taxonomy.*

*The **text** provides information on the appearance of the whale, dolphin, or porpoise; its common names; habitat; and where and what to look for when you want to find it.*

Secondary photographs highlight a distinctive behavior you may observe, or details of particular features that help identify the species. In some cases, part of the cetacean's habitat is shown.

This **illustrated banding** at the top of the spread indicates that it is about a cetacean species.

This **panel** refers to the cetacean family to which the species belongs.

Delphinidae: Oceanic Dolphins

Delphinidae: Oceanic Dolphins

Delphinidae: Oceanic Dolphins

Short-beaked Common Dolphin
Delphinus delphis

ne of the most common of all dolphins, this arly hin is easily d by the hourglass side. This makes a e below the dorsal a reflection of the or another, more ave been proposed variations of this despread species. one other species has lly accepted (see below). behavior that follows applies

FIELD NOTES
■ Male: 5½–7' (1.7–2.2 m); female: 5–6½' (1.5–2 m); at birth: 2½' (0.7 m+)
■ Small fish and cephalopods
■ Mainly offshore
■ Temperate and tropical waters, including inland seas (Black Sea)
■ Insufficiently known, but taken in many fisheries worldwide

to both forms; specific anatomical differences are dealt with in the entry below.

Common dolphins travel in groups of 10 to 500 in most areas, with up to 2,000 or more in the eastern tropical Pacific. These herds are so acrobatic and boisterous that the noise of their approach can often be picked up from miles away. Their high-pitched sounds can sometimes be heard as they bow ride.

Although common dolphins are killed in significant numbers in tuna and other fisheries,

beaked Common Dolphin
Delphinus capensis

FIELD NOTES
■ Male: 6½–8½' (2–2.6 m); female: 6–7½' (1.9–2.3 m); at birth: 2⅓' (0.7 m+)
■ Various small fish and cephalopods
■ Coastal
■ Temperate and tropical world waters
■ Insufficiently known

nsis." Based fferences in at least n within

s a ween its nger,

short-beaked (above) and long-beaked

less chunky body; a less rounded head; and a thicker, dark line between the beak and the flipper. But the most noticeable is that the long-beaked form has a longer beak than the short-beaked form.

Little is known about this dolphin's behavior at sea, apart from what is generally known about common dolphins. However, because of its strong coastal presence, it makes up a sizable proportion of the sightings of common dolphins. 177

Long-finned Pilot Whale
Globicephala melas

FIELD NOTES
■ Male: 13–25' (4–7.6 m); female: 10–18½' (3–5.6 m); at birth: 6' (1.9 m)
■ Squid
■ Offshore
■ Cold temperate to subpolar North Atlantic and Southern Ocean
■ Insufficiently known, still hunted in North Pacific, still hunted in North Atlantic

oth species of pilot whale are sometimes called "potheads," a name that given, as many whale names e, by the hunters who first ountered them. The name is ed on the resemblance of the ot whale's head to a black on cauldron, or pot. The ong-finned pilot whale is ill hunted persistently n the North Atlantic Ocean (see p. 47).

Like the short-finned is pilot whale, the long-finned is a family animal, traveling in groups of 10 to 50, and sometimes up to 100 or more. There are reports of thousands seen together in great superpods. Any sighting of a group of black, medium-sized whales with rounded heads and very wide, thick, curved-back dorsal fins, is bound to be of pilot whales. Young long-finned pilot whales will sometimes breach, but this is rarely observed in an adult.

In a few temperate waters of the world, the distribution of short-finned and long-finned pilot whales overlaps, making it difficult to tell the two apart. Externally they look alike, except that the long-finned has longer flippers (pectoral

fins) and a few more teeth. It is difficult, however, to see flippers, much less teeth, at sea. At times, positive identification may be impossible. It's no wonder that mariners usually write "pilot whales" in their logbooks, refusing to try to distinguish between the two. But in most areas, distribution is quite enough to determine the species.

Long-finned pilot whales inhabit cold temperate to subpolar waters in both the Northern and Southern Hemisphere, except in the North Pacific. The species lived in the North Pacific off Japan until at least the tenth century, but has since completely disappeared.

The long-finned pilot whale often strands on beaches and is possibly subject to more mass strandings than any other cetacean. Frequent strandings occur on Cape Cod, in southern Australia, New Zealand, and southern South America. Because of their social ties, when one pilot whale strands, the others remain with it. 189

long flipper

short flipper

Quick-reference Field Notes panel
■ Size from sexual maturity to maximum length of male and female; average size of calf at birth
■ Main food
■ Habitat
■ Range of the cetacean
■ Status: IUCN category and other notes

Color illustrations supplement the text by showing basic features of the particular whale, dolphin, or porpoise.

Gray Whale

Eschrichtius robustus

Whalers called gray whales "devilfish," mainly because of the ferocity of the mothers when separated from their calves. Whale watchers are still advised to avoid coming between mothers and calves. Young gray whales, however, can be "friendly," coming to the side of boats and even lifting them partly out of the water.

Mature gray whales carry lice and barnacles, as well as numerous scars and marks. Young gray whales are darker than the adults, and have no barnacles or lice. Adults have robust bodies, bulkier than rorqual whales, but much slimmer than right whales. The dorsal fin, only a small hump, is followed by 6 to 12 knuckles extending down the back to the flukes. Gray whales are often active at the surface, spyhopping, breaching, and at times lobtailing and surfing in shallow water.

The gray whale is a messy eater and is the only whale that is known to feed often in the sand and mud, sucking up benthic amphipods, as well as considerable water, sand, and stones.

FIELD NOTES

■ Male: 36½–48' (11.1–14.6 m); female: 38½–49' (11.7–14.9 m); at birth: 15' (4.6 m)

■ Various benthic amphipods, polychaetes, isopods, tube-worms

■ Inshore to open ocean

■ Coastal North Pacific

■ Not listed on IUCN Status list; was intensively whaled, Californian stock is apparently healthy, but Korean stock remains low

For some reason, most gray whales feed by rolling onto their right side, but a few are "left-handed" feeders.

Each year, gray whales in the eastern North Pacific make the 12,400 mile (20,000 km) round trip between Mexico and Alaska. The cruising speed on migration is about 1 to 3 miles (1.6 to 4.8 km) per hour. A few, mainly younger whales, make a shorter journey from Mexico, stopping off in the area from northern California to British Columbia.

With its blotchy complexion, often dirty face, and chunky profile, the gray whale might seem an unlikely Californian hero. Yet everyone respects a survivor. Eastern Pacific gray whales, nearly extinct in the late nineteenth century, are now thought to number more than 20,000.

Gray whales, a characteristic mottled gray and easily identified, are frequently spotted close to shore while migrating.

Northern Right Whale

Eubalaena glacialis

Right whales were so called because the early Basque whalers considered them the "right" whales to hunt—full of oil and easy to catch. The northern right whale is the most endangered whale in the ocean. Its numbers have been reduced to about 300 in the North Atlantic, and possibly a handful in the North Pacific.

A large, bulky animal, the northern right whale is mostly black with distinctive patches called "callosities" on its head. These patches are covered in whale lice—cyamid crustaceans—which make them appear whitish yellow, orange, or pink. The largest patches, above the eyes and around the tip of the rostrum, are arranged in unique patterns that are used to identify and keep track of individuals.

Northern right whales are unlikely to be confused with other animals since they are the only large whales within their temperate range without a dorsal fin and with callosities on the head. But even the clear sight of a flipper or

FIELD NOTES

■ Male: 49–54' (14.9–16.4 m); female: 51–60' (15.5–18.3 m); at birth: 16'+ (4.9 m+)

■ Copepods and other zooplankton

■ Inshore and offshore

■ Temperate and subpolar waters of the Northern Hemisphere

■ Endangered

a fluke is enough for most whale-watch skippers or naturalists to identify them. The flippers are broad and the spatulate shape is unique; the flukes are wide, with a smooth, concave trailing edge and pointed tips, and a deep median notch.

Northern right whales sometimes breach or roll around in the water. When females are ready to mate, they call for males then turn on their backs to play "hard to get" as males charge in from various directions. The males jostle to stay next to the female, waiting for her to roll over to breathe. Then they try to mate. Such behavior can continue for hours.

Once victimized by whalers, the right whale is today a victim of its feeding habits. In some prime right whale areas, shipping traffic needs to slow down to avoid slow-moving right whales as they sift copepods from the water.

The distinctive callosities are clearly visible on these right whales nuzzling each other.

147

Southern Right Whale

Eubalaena australis

Southern right whales have healthier numbers than their northern relatives, but they are still considered "vulnerable" because of intensive whaling. Protected since 1935, along with the northern right whale, the southern right whale (see also p. 96) has been recovering slowly, but probably numbers no more than 3,000 to 5,000.

The southern right whale (as well as the northern right whale) has unique callosities on its head, diagnostic of the species as well as individuals. In 1970, Roger Payne and colleagues in Argentina began studies on the southern right whale, among the first photo-identification studies of whales in the world.

Southern right whales and northern right whales (see p. 147) look almost identical, but are considered to be different species. There are small cranial variations between the two species, and they are reproductively isolated. When the southern right

FIELD NOTES

- Male: <50' (<15 m); female: <54' (<16.4 m); at birth: 15'+ (4.6 m+)
- Copepods, euphausiids and other zooplankton
- Inshore and offshore, mainly coastal
- Temperate and subpolar waters of the Southern Hemisphere
- Vulnerable, was intensively whaled

is on its calving and nursing grounds, the northern right is half a hemisphere away feeding. Some researchers have noted that southern right whales have more callosities on top of their lower "lip" and fewer on the head than the northern species.

Southern right whales engage in behavior known as sailing—raising their flukes into the wind and using them as sails. This behavior appears to be a form of play—although it is never seen among northern right whales—and is a popular activity in the strong, steady winds off Argentina.

In the austral winter months, southern right whales mate and calve in the inshore waters of Chile, Argentina, Brazil, South Africa, southern Australia, and some Southern Hemisphere islands. Most of them migrate to remote southern waters nearer Antarctica to feed during the austral summer. Less is known about their habits and where they go during the main feeding season.

A southern right whale on its breeding grounds, off Argentina.

Bowhead Whale

Balaena mysticetus

The bowhead spends its life following the advance and retreat of Arctic ice. For the long winter, it lives in total darkness in cold waters. In the short summer, the 24-hour sunshine brings an explosion of life and a world bathed in light.

Bowheads are very rotund, with huge heads containing the longest baleen of any whale. Each side of the jaw contains 250–350 plates which grow up to 10–13 feet (3–4 m), and rarely, reaching 17 feet (5.2 m).

Within their home range, size and bulk alone make bowheads distinctive. The arctic toothed whales, the narwhal and beluga, are both much smaller. Compared to gray, right, or any of the rorqual whales, bowheads have smooth skin and are black all over, except for white to gray areas on the tail stock and around a "neck-lace" of black spots on the chin.

The few whale watchers who get a chance to meet this whale generally see only two slightly raised humps on the water. The first is the head, followed by the deep indentation behind the blowhole and then the broad, rounded, black

FIELD NOTES

- Male: 39½–59' (12–18 m); female: 46–65½' (14–20 m); at birth: 11½'+ (3.5 m+)
- Euphausiids, copepods, swimming mollusks, and sea jellies
- Near pack ice and along leads
- Cold arctic and subarctic waters
- Vulnerable, was intensively whaled, still hunted off Alaska, Canada, and Russia

back with no dorsal fin.

Bowheads travel in groups of three or fewer, but form larger groups on the feeding grounds. Despite their size and slow-moving nature, they do breach and often raise their tail flukes before diving. They make frequent low sounds, especially while migrating, and may passively echolocate by listening to the echoes of their own sounds off the ice.

Bowhead whales were the second main species—after the right whale—to be targeted by the commercial whaling industry. Driven to commercial extinction, today only a few are taken each year off Alaska, Canada, and Russia by Inuit and other native people.

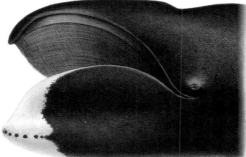

baleen plates of the bowhead whale

Minke Whale

Balaenoptera acutorostrata

The minke whale was formerly called the little piked whale or little finner in some parts of the world. The smallest and most abundant of the rorqual whales, it is found in all oceans, almost to the edge of the ice. Minkes are the only baleen whales still being regularly hunted by commercial whalers. Estimates of the population are therefore controversial, but they range from 500,000 to 1,000,000 individuals worldwide.

A slim whale with a sharply pointed head, the minke surfaces, blows, and rolls through the water, its dorsal fin appearing briefly before it dives. The minke's smaller size (slightly larger than an orca), compared to other baleen whales,

FIELD NOTES
- Male: 22–32' (6.7–9.8 m); female: 24–35' (7.3–10.7 m); at birth: 8'+ (2.4 m+)
- Schooling fish and various invertebrates
- Inshore to offshore
- Cold polar to tropical world waters
- Insufficiently known, currently being hunted

A minke at the edge of the ice in Antarctica.

provides the first indication of the species. A good way to identify a minke is to look for the white band on the flippers. This is clearly visible at medium to close range as a white patch just below the surface beside the animal. In some parts of the world, however, the flippers are all black.

When seen up close, the back of the minke whale is swathed in variable areas of dark and light pigmentation. These marks have been used to photo-identify and keep track of individuals.

On their feeding grounds minkes are often seen in pairs, although the numbers can reach up to 100 in a feeding area where food is plentiful. Minkes tend to feed steadily with little concern for boats or people, but sometimes, in certain places, they can exhibit curiosity around boats, even swimming beside a boat for 30 minutes or more.

minke flipper banding

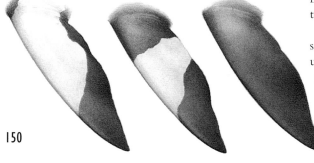

Bryde's Whale

Balaenoptera edeni

Bryde's whales are unique among baleen whales because they spend the entire year in the tropical and subtropical zones, in waters more than 68° Fahrenheit (20° C). They make only short migrations—or none at all—and never visit cold waters. Sometimes called the tropical whale, the Bryde's (pronounced broo-dess) whale is named after Mr. J. Bryde, who helped build the first whaling factory in Durban, South Africa, in 1909.

Smaller than the sei whale, Bryde's whale is slender with short flippers, a prominent, curved-back dorsal fin, and the classic rorqual head— narrow with a pointed snout and numerous throat grooves that enable the mouth to expand when feeding. It is dark above and light on the throat and belly, and the skin is sometimes spotted with circular scars. The most distinguishing characteristic is the set of three parallel ridges

FIELD NOTES

■ Male: 39–47' (12–14 m); female: 41–51' (12.5–15.5 m); at birth: 11' (3.4 m)
■ Fish and some invertebrates
■ Inshore to offshore
■ Tropical and subtropical world waters
■ Insufficiently known, has been whaled to some extent

on the top of the head extending from the tip of the beak to the blowhole. Sei and other baleen whales have only one ridge down the center.

Typical feeding behavior can be observed at two key sites. In the Gulf of California, Mexico (above), the resident whales feed alone, or may be found in groups of up to five or more, although they are usually widely separated on the feeding grounds. In Tosa Bay, Kochi Prefecture, Japan, the 15 or more resident Bryde's whales feed alone or in mother-calf pairs. Commonly seen here when the animal breaches is the pinkish tinge to the throat and belly. Bryde's whales in both areas will approach whale-watch boats.

There may be separate inshore and offshore populations in various parts of the world. As photo-identification studies are fairly recent, much has yet to be learned.

Bryde's whale head

sei whale head

Sei Whale

Balaenoptera borealis

T he sei is the middle whale in size in the series of Balaenopterid rorquals (from smallest to largest: minke, Bryde's, sei, fin, blue). It looks like a typical rorqual, with the characteristic narrow snout; the erect, curved-back dorsal fin; and the slender flippers.

At a distance, sei whales are easily confused with the rorquals closest to it in size: Bryde's whales and fin whales. However, Bryde's whales have three ridges on the top of the head, while the sei whale has a single ridge from the tip of the rostrum to the blowhole (see p. 151). The sei whale is smaller than the fin, and unlike the fin, has symmetrical coloring on either side of its head. You can easily tell a sei from a fin by observing both sides of the whale's head just as it surfaces.

Feeding behavior is varied. Like right whales, sei whales skim for copepods, swimming

FIELD NOTES
- Male: 42–60½' (12.8–18.5 m); female: 44–69' (13.4–21 m); at birth: 15' (4.6 m)
- Euphausiids, copepods, amphipods, small fish, squid
- Offshore
- Tropical to subpolar world waters
- Vulnerable, was intensively whaled

steadily through the water to catch their food, rather than lunging and gulping like other rorquals. For this reason, their ventral pleats are short and the baleen fringes fine. However, in the Southern Hemisphere, krill is the main food, but they will turn to copepods and amphipods. Sometimes great numbers of sei whales follow prey into prime fishing areas. They usually travel in groups of two to five individuals, although on good feeding grounds, many more, often well spaced, are observed.

Sei whales are rarely seen because of their preference for offshore waters. Most of what we know about them comes from studies associated with whaling or whaling management. After the commercial whaling industry ran out of blue and fin whales, whalers turned to sei, as well as Bryde's whales. Sei whales soon became scarce, but they may have increased in recent years because of whaling restrictions.

blue

fin

sei

Bryde's

minke

Fin Whale

Balaenoptera physalus

FIELD NOTES

- Male: 58–82' (17.7–25 m); female: 60–88½' (18.3–27 m); at birth: 19½'+ (5.9 m+)
- Schooling fish, euphasiids and other invertebrates, copepods, squid
- Inshore to mostly offshore
- Warm temperate to polar world waters (usually migratory)
- Vulnerable, was intensively whaled

Individual fin whales are identified by the V-shaped chevron behind the head, the back pigmentation, and dorsal fin marks.

The second-largest living animal after the blue whale (see p. 154), the fin whale's record length is about 88½ feet (27 m). Because of its size and cosmopolitan presence, it was extensively whaled from the late 1800s in the North Atlantic, then in the North Pacific and the Southern Ocean. By the 1930s, when blue whale numbers began to decline, fin whale catches had increased in the Southern Ocean. In the 1970s, seriously declining stocks were protected from whaling, although some commercial whaling continued in the North Atlantic until 1985. Native whaling still occurs off Greenland.

At sea, the fin whale is most likely to be confused with blue and sei whales. All others are much smaller. However, the fin whale has a dramatic asymmetry which whale watchers can easily observe: white on the lower right side of the head; black on the lower left. Sometimes, if the whale is feeding, you can also see the mostly white baleen on the right side; the baleen on the left side is dark gray.

Fin whales are one of the more commonly seen whales in the Northern Hemisphere—they can be seen close to shore off Iceland, eastern Canada, New England, Baja California, and in the Mediterranean. They can also be encountered, although less often and perhaps a little farther offshore, in the Southern Hemisphere, including in the Antarctic.

Fin whales rarely breach or spyhop when being observed and normally just keep feeding. They are often observed alone or in pairs, but on feeding grounds up to 10 or 20 can be found together, with 100 or more at times loosely grouped. In the St. Lawrence, they often travel in pods of 5 to 10, swimming and feeding together in a tight group.

A photo–identification catalog has been set up for the North Atlantic and many researchers are contributing their photos and sightings in an effort to learn more about these animals.

Blue Whale

Balaenoptera musculus

Almost every aspect of the blue whale's appearance and natural history provides another footnote for the record books. This is the largest living animal, but the largest known blue is subject to debate. The "best" record appears to be at least 110 feet (33.6 m) long and a weight of 209 tons (190 tonnes). It was a female—the largest baleen whales are females.

A newborn blue whale measures at least 19½ feet (5.9 m). Biologists have made estimates about its first year of life: a baby blue whale drinks 50 gallons (190 L) of milk a day, adding about 8 pounds (3.6 kg) of weight per hour, or 200 pounds (90 kg) a day. At about eight months of age, when the calf is weaned, it can measure close to 50 feet (15 m) long and weigh about 50,000 pounds (22,700 kg).

Blue whales are so much larger than other whales that distinguishing them is usually easy.

FIELD NOTES

- Male: 65½'–101½' (20–31 m); female: 69–110' (21–33.6 m); at birth: 19½'+ (5.9 m+)
- Euphausiids with some squid, amphipods, copepods, red crabs
- Offshore
- Tropical to polar world waters (migratory)
- Endangered, was intensively whaled

A blue whale gulps a big mouthful of food and water.

Compared to the next largest, the fin whale, the blue whale's spout is thicker and taller. If you cannot see the fin whale's white markings on the right-hand side of the head, look for the shape of the head. Seen from above, a blue has a wide, almost U-shaped rostrum, whereas the fin whale has a more V-shaped head with a pointed snout. In a certain light, it is possible to see the characteristic mottling of the blue whale's back and sides. The mottling picks up reflected blues of the sea and sky, which is the origin of its common name. Individual whales can be identified by the blotchy pigmentation.

At sea, blue whales usually feed alone or in pairs, often widely spaced, probably because they need to work large areas. Fast-moving for their size, blue whales are thought to be able to exceed 19 miles (30 km) per hour. In the late nineteenth century, they were pursued relentlessly; by the 1950s, blue whales were endangered. Only an estimated 6,000 to 14,000 remain in the world's oceans.

Humpback Whale

Megaptera novaeangliae

The most popular whale on whale-watch trips, the humpback seems happy to perform: it lifts its head out of the water, waves its long, massive flippers, splashes its tail, rolls over in the water, and more than anything else, leaps out of the water. It tends to move close to boats, as if attracted by watchers, and shows little shyness toward humans. There are more spectacular pictures of humpbacks than of any other whale.

The humpback whale is classed as one of the rorquals, but some aspects of its anatomy differ. Compared to the slim, sleek, classic rorquals (see p. 152), the humpback is bulkier, and its skin is knobby and covered in barnacles. The dorsal fin is reduced to a fleshy hump, or hook, that sits on a sort of platform on the back. The tail is not smooth but ragged on the trailing edge. More than anything else, the flippers seem greatly out of proportion. At up to 16 feet (5 m) long, they are the longest appendages of any animal.

Although said to be slow-moving, humpbacks are capable of impressive bursts of speed during mating and fighting. Their feeding behavior is more versatile than other baleen whales and includes bubblenetting (see p. 67).

Humpbacks are seen singly, in pairs, or in groups of up to 15 animals. Mothers and calves stay close for the first year, but on the mating grounds they are often joined by male "escorts."

Humpbacks usually lift their flukes before a deep dive. The flukes have a distinct pattern on the underside, which varies from almost white with black markings, to almost black with only a few white markings. Several thousand humpback individuals have been photo-identified by their fluke markings and have become part of huge catalogs of whales in the North Atlantic, North Pacific, and in the Southern Hemisphere.

FIELD NOTES

- Male: 36–57½' (11–17.5 m); female: 36–62½' (11–19 m); at birth: 15'+ (4,6 m+)
- Various schooling fish, invertebrates
- Mainly inshore, but also offshore
- Tropical to polar world waters
- Vulnerable, was intensively whaled, numbers probably increasing

A humpback surfaces—its mouth full of fish from bubblenetting.   155

Sperm Whale

Physeter macrocephalus

In many ways, the sperm whale can claim the title "lord of the sea." It ranges throughout most of the world's oceans, except the high Arctic. It has the largest brain of any animal, and is the largest toothed whale—the male can be as long as 65½ feet (20 m). It seems fitting that Herman Melville chose the sperm whale for his novel *Moby Dick*.

The sperm whale has a number of physical features that make it almost instantly identifiable. At a distance, first seen is often the spout, which is directed forward and a little to the left (the blowhole is on the left side of the forward portion of the head). At closer range, you can see the low hump that barely passes for a dorsal fin, followed by a series of bumps leading to the tail.

The skin can be splotchy, with scratches on the head, and the middle to rear parts of the body prune-like. When a sperm whale breaches, the huge head, up to a third of the body length, is visible. Careful observation reveals white areas around the mouth.

FIELD NOTES

- Male: 36–65½' (11–20 m); female: 27–56' (8.2–17 m); at birth: 13' (4 m)
- Deep-water squid, including giant squid and larger fish
- Offshore
- Tropical to subpolar world waters
- Insufficiently known, but still exist in some numbers

Sperm whales spend most of their lives in either nursery or bachelor schools (see p. 94–5). A nursery school includes females of all ages plus immature males. Mature bulls visit during the breeding season. A bachelor school includes males that are sexually mature, or nearly so. Differences between male and female sperm whales are pronounced. Mature males can be one and a half times the length of mature females. Most females have calluses on the dorsal hump, which males rarely do.

Sperm whales make long, deep dives, possibly up to 10,000 feet (3,000 m) that can last up to two hours. Before diving, they lift their flukes. Although never witnessed, their battles with giant squid are the stuff of legend. Scars on the bodies of sperm whales, and giant squid beaks found in the stomachs of dead sperm whales, provide compelling evidence.

Sperm whales usually congregate in nursery or bachelor schools of up to 50 animals.

Dwarf Sperm Whale

Kogia simus

Whale-watch tours from the Gulf of California in Mexico, Dominica in the Caribbean, and the Tañon Strait in the Philippines, have encountered this little-known, dolphin-sized whale repeatedly. In the Caribbean it is often difficult to tell it apart from its close relative, the pygmy sperm whale. These two whales were declared separate species as recently as 1966.

Underwater, and when found stranded on beaches, the heads of both look shark-like. The dwarf sperm whale is a little smaller than the pygmy sperm whale, but the dorsal fin is larger

FIELD NOTES
- Male: 7–9' (2.2–2.7 m); female: 7–9' (2.2–2.7 m); at birth: 3½' (1.1 m)
- Deep-water squid and octopus, fish, invertebrates
- Offshore
- Tropical to temperate world waters
- Insufficiently known

and more erect—somewhat like a bottlenose dolphin's dorsal fin. Either a glimpse of the head, or a few minutes' observation at sea, will confirm that this isn't a dolphin; the challenge is to distinguish a dwarf sperm from a pygmy sperm whale.

When resting, the dwarf sperm may float lower in the water than the pygmy sperm whale. It is slow-moving and drops directly below the surface instead of continuing to swim forward as it dives. It often travels alone or with a companion, although groups of up to 10 are sometimes seen.

Pygmy Sperm Whale

Kogia breviceps

The shark-like head, false gill markings, and under-slung lower jaw make the pygmy sperm whale look like a fish at first glance. At sea, it rises slowly to breathe, then drops. The dorsal fin is small compared to its overall body size and smaller than a bottlenose dolphin's or a dwarf sperm whale's—two cetaceans with which it could be confused. It is difficult to make a positive identification except during rest periods when it floats at the surface, its head and back exposed and only its tail hanging limply in the water. There are usually up to six animals in a group.

A warm-water, offshore species, these whales frequently strand on beaches,

FIELD NOTES
- Male: 9–11' (2.7–3.4 m); female: 8½–9½' (2.6–2.9 m); at birth: 4' (1.2 m)
- Deep-water squid and octopus, fish, invertebrates
- Offshore
- Tropical to temperate world waters
- Insufficiently known

especially in South Africa, New Zealand, southeastern Australia, and on the east coast of North America. Most of what we know about them comes from strandings. Rarely seen in the wild, whale-watch tours off Dominica have encountered them recently. They seem to be shy animals, seldom approaching boats. When frightened, they may release brown fecal material that clouds the water and may work as a decoy, like squid ink.

Pygmy sperm whale

157

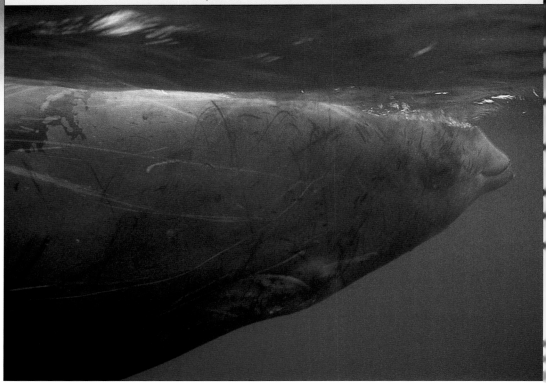

Cuvier's Beaked Whale

Ziphius cavirostris

One of the most abundant and widespread beaked whales in the world, Cuvier's is one of the three most watched beaked species (along with Baird's beaked whale and the northern bottlenose whale).

The first details of this species were published in 1823 when the French anatomist Georges Cuvier created a new genus and described what he thought to be an extinct whale. In the 1870s, it was realized that Cuvier's fossil represented a living species, and numerous disparate beaked whale findings from all over the world were re-identified as Cuvier's beaked whales.

The forehead slopes gently to a slight beak. Because the head, often visible as the animal swims, has a shape reminiscent of a goose's beak, Cuvier's beaked whales are sometimes called goose-beaked whales. Often depicted as brown or black, their color varies from individual to individual. Older males, for example, have extensive

FIELD NOTES

■ Male: 17½–22½' (5.3– 6.9 m); female: 17–21½' (5.1– 6.6 m); at birth: 7'+ (2.2 m+)

■ Deep offshore cephalopods, some fish and crustaceans

■ Offshore

■ Tropical to cold temperate world waters, including Mediterranean Sea

■ Insufficiently known

white areas from the beak to the top center of the body. The whales are heavily scarred—particularly the older males—from the teeth of other males. The two teeth erupt only in males and can be seen protruding from the tip of the lower jaw. Barnacles sometimes grow on the teeth.

The dorsal fin, often the first feature seen at sea, is curved back like that of a dolphin or minke whale, and is positioned far back on the body. Cuvier's beaked whales travel alone (usually older males) or in groups of up to 25, although more commonly 10 or fewer. They arch their back steeply and sometimes raise their tail when diving deep. They may stay down 20 to 40 minutes while hunting for deep-sea fish and squid.

It is difficult to predict sightings, but Cuvier's beaked whales are sometimes seen on whale-watch trips in the Mediterranean, Hawaii, the Canary Islands, and off South America en route to Antarctica.

lower jaw and teeth of Cuvier's beaked whale

158

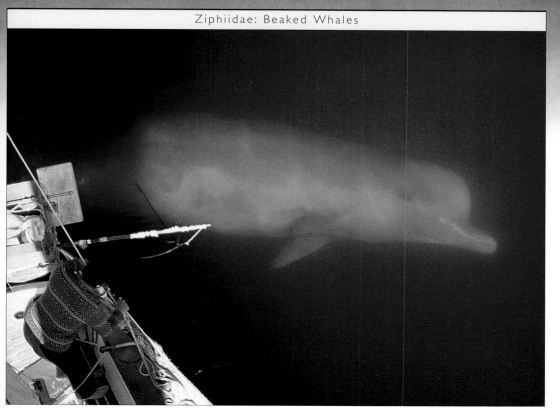

Northern Bottlenose Whale

Hyperoodon ampullatus

T he northern bottlenose whale, particularly an older male, has a large bulbous forehead and a prominent tube-like beak. The female is smaller with a less pronounced forehead and beak. Both sexes regularly approach ships, and this behavior has made them more likely to be hunted, studied, and, more recently, whale-watched than the other beaked whales.

Resident in the cooler waters of the North Atlantic, northern bottlenose whales can be reliably found in the Gully (see p. 215). The whales make dives of 3,280 feet (1,000 m) or more, using their superb sonar system to pursue deep-water squid in the black depths.

Many researchers have watched them playing and socializing; leaping out of the water, and touching one another. Long-term relationships have been observed among whales of the same sex, but not among mixed sexes. Using photo-identification, the researchers have shown that

FIELD NOTES

■ Male: 24–32' (7.3–9.8 m); female: 19–28½' (5.8–8.7 m); at birth: 11½' (3.5 m)
■ Squid, some fish and invertebrates
■ Deep waters offshore
■ Cold temperate waters of the North Atlantic
■ Insufficiently known, heavily whaled but healthy population in the Gully

about 230 northern bottlenose whales use the Gully, and that they do not make seasonal migrations. They are among the few whales known to spend the entire year in cold waters.

Northern bottlenose whales have been hunted more than any other beaked whale. In the late nineteenth century, Scottish whalers used to call the old northern bottlenose whale bulls "flatheads." Tens of thousands have been killed since then. Scientists disagree about the extent to which the species has been depleted, and its current status. Increased awareness of northern bottlenose whales through responsible whale watching in the Gully and other parts of their range, may help secure the future for this "friendly" and attractive species.

The bulbous foreheads of northern bottle-nose whales protrude above the surface.

159

Baird's Beaked Whale

Berardius bairdii

Baird's beaked whales inhabit the deep offshore waters of the northern North Pacific. The species was discovered in 1882 when researcher Leonhard Stejneger picked up a four-toothed skull on Bering Island. The following year, he published his discovery, honoring his colleague, Spencer Baird, with the species name. Baird had also worked in Alaska and had just been appointed Secretary of the Smithsonian Institution.

Yet long before this large, up-to-42 foot (12.8 m) beaked whale was classified, the Japanese were catching it with hand harpoons as part of coastal whaling. Today, the Japanese still capture 40 to 60 annually, mainly in the waters off the Boso Peninsula, near Tokyo, and off Hokkaido Island in the north.

Baird's is probably the largest of the beaked whales, with a long beak like a bottlenose dolphin. Unlike most other beaked whales, the

FIELD NOTES

- Male: 30–39' (9.1–12 m); female: 32–42' (9.8–12.8 m); at birth: 15' (4.6 m)
- Deep-sea cephalopods, crustaceans, and fish
- Offshore
- Warm to cold temperate waters of the North Pacific
- Insufficiently known, still hunted off Japan

lower jaw and teeth

males and females both have teeth that erupt: one large pair at the tip of the protruding lower jaw and a smaller pair just behind them. In older animals, the teeth can be worn down to the gums. The bulging forehead is broader and more bulbous in males than females, although females on average are larger. Baird's beaked whales stay in tight social groups ranging from 3 to 30 or more individuals.

Like other beaked whales, Baird's beaked whales dive deeply. Dive times can reach up to 67 minutes, but 25 to 35 minutes is more common. The scarring on their back indicates that there is probably a great deal of play or aggression within the groups.

The chance to see this species in the wild affords a rare opportunity to meet a beaked whale in action. Even for a seasoned whale watcher, who may have met orcas, gray, humpback, right, and sperm whales, a beaked whale encounter would be something special.

Blainville's Beaked Whale

Mesoplodon densirostris

I n 1972, a young male Blainville's beaked whale came ashore on a New Jersey beach. It was alive, and James G. Mead from the Smithsonian Institution drove to the scene and tried to save it. The whale lived for three days, and died without revealing much. Mead has since worked hard to learn more about the mysterious beaked whales.

Blainville's beaked whale was one of the early beaked whales to be identified. In 1817, Henri de Blainville managed to describe it from a small piece of the jaw, which was the heaviest bone structure he had ever seen, denser even than elephant ivory. This also gave rise to its other common name: the dense-beaked whale.

The Blainville's beaked whale forehead is flat and the dorsal fin is triangular or curved back. One of the beaked whales with a high-arched jawline, the male has two large teeth on the crest of the arch that grow forward and protrude from both sides of the closed

FIELD NOTES
- Male: <19 ½' (<5.9 m); female: <15½' (<4.7 m); at birth: 6½'+ (2 m+)
- Squid and other cephalopods
- Offshore
- Temperate to tropical deep ocean waters
- Insufficiently known

female (above) and male with erupted tooth in lower jaw

mouth. The female has a beak lighter in color, a less prominently arched lower jaw and, as with most other beaked whales, no erupted teeth.

Traveling in groups of up to 6 (occasionally up to 12), Blainville's beaked whales are often heavily scarred, particularly the males. Circular scars probably come from cookie-cutter sharks and various parasites, while the long, white lines may come from fights with other males. Their deep dives for food can last for up to 45 minutes.

This whale turns out to be one of the more widely distributed beaked whales, commonly stranding on oceanic islands on both sides of the equator. It has been seen on whale-watch trips off Oahu in Hawaii, where researchers were able to photograph the male's head as it surfaced headfirst at a sharp angle, the sun gleaming on the barnacles that covered the two large teeth.

161

Sowerby's Beaked Whale

Mesoplodon bidens

More than one naturalist has called Sowerby's the "North Sea beaked whale." Although its range extends across the North Atlantic, this whale is most likely to be found in the northern North Sea. It was first discovered off north-eastern Scotland in 1800. A male had stranded in the Moray Firth and the skull was collected. A couple of years later, James Sowerby, an English watercolor artist, painted a picture of the skull and how he imagined the animal looked. When the pictures were published, he used the Latin name *bidens*, meaning "two teeth," to name the species. Because this was the first beaked whale ever described, the teeth seemed unusual but, in fact, although the position often varies, the males of many species of beaked whale have two pro-truding teeth in the lower jaw.

The two teeth in the Sowerby's male are located about halfway along the fairly straight mouthline of the lower jaw and are visible even when the mouth is closed. If you could approach close enough at sea,

FIELD NOTES
- Male: 18' (5.5 m); female: 16½' (5 m); at birth: 8' (2.4 m)
- Squid and small fish
- Offshore
- Cold temperate North Atlantic, including North Sea
- Insufficiently known

this would be the best way to distinguish them from other beaked whales. Compared to other whales found in the North Atlantic, Sowerby's has a smaller head, longer dolphin-like beak, smaller dorsal fin, and flukes with no median notch. It is smaller than the minke whale.

Although Sowerby's was the first beaked whale to be named, it remains one of the more elusive. It has rarely been seen alive at sea. Although a few have stranded on beaches, basic natural history data, such as the contents of the stomach, have not been recorded.

In the early 1970s, a calf and its dying mother were found on a beach in Belgium. A Dutch aquarium tried to rehabilitate the calf, which was 9 feet (2.7 m) long and weighed 408 pounds (185 kg). This provided a chance to see how the animal swam, tucking its flippers into its side and doing all the propulsion and steering with its tail (dolphins use their flippers to steer). Its lack of fine steering ability led to the animal's death as it hit the walls of the small pool.

lower jaw and tooth of Sowerby's male

Strap-toothed Whale

Mesoplodon layardii

The strap-toothed whale presents one of the puzzles of the whale world. The two teeth in the lower jaw of the male grow up and over the upper jaw, sometimes wrapping right around it. In this position, the jaw can open only a little way, so how does the male strap-toothed whale catch food, much less swallow it? Perhaps it sucks up its food (squid) from the tight corners of deep offshore canyons. The teeth are not needed for chewing—the female survives with no erupted teeth. Some researchers claim that "wrap-around" teeth may function as guard rails to funnel food into the mouth. Most likely the teeth are a sexual characteristic, and help determine the fittest males for mating. The variety of tooth formations originally led to the naming of at least four new species, but in time, scientists recognized that they were all the same species.

At up to 20 feet (6 m) long, the strap-toothed whale is one of the largest beaked whales. At home in the cold temperate Southern Hemisphere, it has turned up on the

FIELD NOTES

- Male: 19' (5.8 m); female: 20' (6 m); at birth: 7' (2.2 m)
- Squid
- Offshore
- Cold temperate circumpolar waters in the Southern Hemisphere
- Insufficiently known

beaches of Chile and Argentina including Tierra del Fuego, southern Africa, New Zealand, and Australia.

The Southern Hemisphere has produced three new species of beaked whales since the early 1990s. The pygmy beaked whale, *Mesoplodon peruvianus*, up to 12 feet (3.7 m) long, was found in fish markets and on beaches in Peru. The smallest of the beaked whales, it was officially named in 1991. Bahamonde's beaked whale, described in 1995, turned up in the Juan Fernández Islands, 800 miles (500 km) off the coast of Chile. A species called simply "unidentified beaked whale A" has been seen at least 30 times in the eastern tropical Pacific, but it has never been found stranded. A fresh male specimen will be needed to confer full species status and to give it a name.

Beaked whale authority James Mead, of the Smithsonian Institution, has said there may be more beaked whale species still to be discovered. Keep a lookout on deep-sea boat trips: you may be the one to encounter a whale no one has ever seen before.

wrap-around teeth of male

163

Baiji

Lipotes vexillifer

Only 80 years after it was first described in scientific literature, the Yangtze River dolphin, or baiji, is on the brink of extinction. Despite more than a decade of international conservation efforts, the baiji remains the most endangered of cetaceans.

It has a stocky body about the size of an adult human. Like other river dolphins, it has tiny eyes and a long, narrow beak. Its coloring appears white or gray at a distance, but close up looks dark bluish gray on the back fading to grayish white on the belly. The triangular dorsal fin is set low, and the flippers are broad and somewhat rounded.

It is most active from early evening to early morning. People who have been lucky enough to find baiji see them alone or in groups of up to six where tributaries join rivers, especially around shallow sandbanks. The species is quiet, reserved, and difficult to approach. In calm conditions, the blow may be heard as a high-pitched sneeze, although this quick spouting sound is

FIELD NOTES

- Male: 6½–7½' (2–2.3 m); female: 6½–8' (2–2.4 m); at birth: 3' (0.9 m)
- Small freshwater fish
- Yangtze River in China
- Endangered, may soon be extinct

Baiji is found only in the dark blue area of the Yangtze River, as shown above.

difficult to distinguish from that of the finless porpoise, the only other cetacean species in its range. Cetacean sightings in the Yangtze River often turn out to be finless porpoises, because these are more numerous, more visible, and easier to approach.

The baiji was declared a National Treasure of China and has been protected since 1975. Parts of the river have been declared a natural reserve, but this initiative has had little real success in protecting the baiji because of continual boat traffic, fishing, and industrial development along what is one of the world's busiest waterways. The baiji's genus name *Lipotes* comes from the Greek meaning to be "left behind." This refers to its limited range and may yet prove to be its epitaph.

A baiji stamp (right) issued by the Chinese government to win support for conservation.

Boto

Inia geoffrensis

The boto, the largest and most commonly seen river dolphin, is the only one of the five river dolphin species to support regular commercial tours. It is noted for its pink coloring and is variously called the pink porpoise, pink dolphin, or the Amazon River dolphin. The color, however, varies according to sighting conditions, the animal's age, and individual variation. Some botos are bluish gray, while others appear almost white; younger animals are gray.

The only other dolphin to share its habitat in the Amazon and Orinoco regions, is the tucuxi which looks like a small bottlenose dolphin. The boto is larger, with a long beak, broad flippers, a bulbous melon, and a hump instead of a dorsal

FIELD NOTES

- Male: 6½–9' (2–2.7 m); female: 5–7½' (1.5–2.3 m); at birth: 2½' (0.7 m)
- At least 50 species of fish
- Tropical river
- Amazon and Orinoco river basins
- Vulnerable

fin. Its physique is almost ungainly. The large cheeks are so chubby that they may even hamper the dolphin's vision when it is hunting, and this may be the reason that it often swims upside down.

Botos are said to be more active in the early morning and late afternoon, and if so, they follow the pattern of many tropical rain-forest creatures. They are usually seen alone or in pairs, although groups of 10 or 15 may be observed in the dry season when the river levels are low, or if there is plenty of food.

The boto has the widest distribution and largest population of all the river dolphins. It is less affected by human population pressures than the three river dolphins of Asia, but the destruction of the tropical rain forest is causing boto numbers to decline. Conservation lessons learned from the more endangered Asian river dolphins are being applied to the boto, but only time will tell if it has a long-term future.

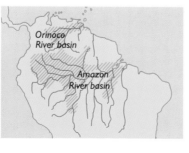

The mouth (left) is fitted with 46 to 70 teeth. Boto's range (above) covers a wide area.

Tucuxi

Sotalia fluviatilis

Tucuxi, the "other dolphin" of the Amazon and Orinoco, is smaller and more plainly colored than the pinkish boto, or Amazon River dolphin (see p. 165). Pronounced too-koo-she, the tucuxi is a "river dolphin" by habit but not a member of the group of five river dolphin species that have evolved to live only in rivers. Although some tucuxi dolphins spend their entire lives in the river, many more live along the coastal area of the western Atlantic from southern Brazil and north, to Panama. These two separate populations vary only slightly, although the coastal form grows larger than the riverine form. One of the smallest dolphins, the chunky tucuxi resembles the bottlenose dolphin in overall body shape. Along the coast, it may be confused with the bottlenose, but the tucuxi is much smaller, with a smaller triangular dorsal fin.

Watching the larger coastal
tucuxi, off Rio de Janeiro.

FIELD NOTES

- Male: <7' (<2.2 m); female: 4 ½–7' (1.4–2.2 m); at birth: 2'+ (0.6 m+)
- Variety of fish, mainly schooling; cephalopods
- Inshore shallow coast and rivers
- Tropical waters of the western Atlantic Ocean, including the Amazon and Orinoco rivers
- Insufficiently known, killed accidentally in nets

Tucuxi travel in groups of 2 to 7, although larger groups may consist of as many as 20 in fresh water and 50 along the coast. They engage in typical dolphin behavior, leaping, spyhopping, flipper slapping, and lobtailing, but do not bow ride. Along the coast of Brazil, the sight of a tucuxi leaping high out of the water then crashing back on its side is a familiar one. It then rejoins its family and swims on.

For years, in Brazil, the tucuxi has been killed in purse-seine and gill nets, as well as shrimp traps. Recently, Brazilian researchers have begun to study the coastal form in southeastern Brazil, photo-identifying individuals and following them from year to year. The most popular dolphin-watch tours in South America are to see the tucuxi living around Santa Catarina Island, in Brazil. Every week, thousands of people go dolphin watching there on sail and motorboats.

Indo-Pacific Hump-backed Dolphin

Sousa chinensis

In 1765, Swedish explorer Per Osbeck was amazed to find "snow white dolphins at play in the China Sea." In recent years, the "pink dolphins" of Hong Kong have become a great attraction, but just as tours became popular in the mid-1990s, Hong Kong's new airport threatens to destroy the hump-backed dolphins' locally favored habitat.

The Indo-Pacific hump-backed dolphin can often be identified purely by its surfacing behavior (see Atlantic hump-backed dolphin below). It lives in the shallow coastal waters of southern and eastern Africa, including the Red Sea, extending east along the coast to China, the Indonesian archipelago and northern

FIELD NOTES

- Male: <10 ½' (<3.2 m); female: <8' (<2.4 m); at birth: 3 ½' (1.1 m)
- Fish, crustaceans, and mollusks
- Inshore coastal, estuaries and mangroves
- Warm temperate to tropical waters of the Indian Ocean and western Pacific
- Insufficiently known

Australia. Its taxonomy is still being argued, and there may be two species, or at least two distinct populations. Those west of Sumatra have a fatty hump or platform below the dorsal fin, while those east and south of Sumatra have no platform but a taller dorsal fin. The color seems to vary locally, as well as among age groups and individuals. The back is mainly dark gray, but can be almost white or even pink. On the underside, they are light gray.

Indo-Pacific hump-backed dolphins often swim with other dolphins, mainly bottlenose, but also spinner dolphins and finless porpoises. They rarely bow ride or approach boats.

Atlantic Hump-backed Dolphin

Sousa teuszii

For as long as local people can remember, the mullet fishers of Mauritania have waited for Atlantic hump-backed dolphins to help them drive fish into their nets. Resident only in shallow coastal waters of West Africa, these dolphins are often found in the estuaries and mouths of rivers and around mangrove swamps.

Juveniles, darker in color and without a hump, breach frequently and adults chase each other around at high speed, perhaps displaying courtship behavior. Atlantic hump-backed dolphins sometimes associate with

FIELD NOTES

- Male: 6 ½–8' (2–2.4 m); female: <7 ½' (<2.3 m); at birth: 3' (0.9 m)
- Mullet and other schooling fish
- Inshore coastal, estuaries and mangroves
- Subtropical to tropical waters of the eastern Atlantic (west African coast)
- Insufficiently known

Atlantic hump-backed dolphin

bottlenose dolphins in the herding of fish. They are closely related to the Indo-Pacific hump-backed dolphins, but their ranges do not overlap and they usually have fewer teeth.

Both hump-backed dolphin species present an odd sight as they swim in groups of up to 7 and rarely more than 25 individuals. Each animal first lifts its long, slender beak out of the water. There is a clear glimpse of its big torso as it arches its back, displaying the small dorsal fin that sits on a raised platform. The animal then appears to pause—unlike the surfacing behavior in other dolphins—and either dips below or flips its tail and dives.

167

Rough-toothed Dolphin

Steno bredanensis

From a distance, as it surfaces and moves rapidly along, the rough-toothed dolphin looks like a bottlenose dolphin, or even a spotted or spinner dolphin. It sometimes travels among these other species. However, at close range, a rough-toothed dolphin is more easily identified than others.

The name rough-toothed dolphin comes from the fine, vertical ridges on the teeth, but these are impossible to see in the wild. An ungainly, often scarred, primitive-looking cetacean, the rough-toothed dolphin may be the "ugly duckling" of dolphins. The head is sometimes described as having a reptilian appearance. The large flippers look too big for the animal and are set far back on both sides. Its body often has scratches and scars, and bite marks from cookie-cutter sharks. Only the curved-back dorsal fin looks like that of an "ordinary" dolphin.

The rough-toothed dolphin is rarely seen and studied because of its preference for deep waters beyond the continental shelf. In recent years, however, they have been encountered more often around Hawaii, the Bahamas, and off

FIELD NOTES

- Male: 7–8½' (2.2–2.6 m); female: 7½–8' (2.3–2.4 m); at birth: 3' (0.9 m)
- Offshore cephalopods and fish
- Mainly offshore
- Warm temperate to tropical world waters
- Insufficiently known

Ogasawara, Japan. The rough-toothed dolphin is gregarious, traveling in groups of 10 to 20, and sometimes up to 50 or more at a time. They may bow ride but rarely do a full breach. Instead, when traveling fast, they skim the water with just the head visible—a behavior called porpoising. They often stay at the surface only briefly, perhaps because they are busy feeding at depths on small schooling fish and squid.

Rough-toothed dolphins are sometimes captured by accident and killed in tuna purse-seine and other nets. They have been killed in small numbers around Japan, Sri Lanka, and the Caribbean, but little is known about this species and more work needs to be done.

Rough-toothed dolphins porpoising near the Azores Islands.

Bottlenose Dolphin

Tursiops truncatus

The bottlenose dolphin is the archetypal dolphin, found around the world, from cooler temperate to tropical waters. It lives both inshore and offshore, and is the active dolphin that leaps, bow rides, bodysurfs, splashes its tail, and approaches boats and swimmers more than any other dolphin. The lone, sociable dolphins that have mixed with humans for years, such as Fungie in Ireland and the dolphins of *Flipper* fame, are also mainly of this species.

Bottlenose dolphins are mostly gray with a lighter or white underside, a short but definite beak, and a prominent curved-back dorsal fin. The flippers are pointed. The body is robust, but there is great variation in adult size, from 7½ to 12½ feet (2.3 to 3.8 m). Size and other features vary according to whether the dolphins live inshore or offshore and in what part of the world. Through photo-identification studies, researchers have learned

FIELD NOTES

- Male: 8–12½' (2.4–3.8 m); female: 7½–12' (2.3–3.7 m); at birth: 2½'+ (0.7 m+)
- Fish, cephalopods, invertebrates
- Inshore to offshore
- Temperate to tropical world waters
- Insufficiently known, overall numbers substantial but some populations threatened

Bottlenose dolphins have long associated closely with humans.

that coastal bottlenose dolphins reside in or return to the same areas year after year. Females with calves stay together, using the most productive areas of the community home range; males form long-term bonds with each other and range farther afield as they get older. Sometimes the males venture into the range of nearby bottlenose dolphin communities, traveling from one female group to another.

The food habits and hunting behavior of bottlenose dolphins vary greatly. They adapt their behavior to local circumstances and conditions, and only orcas eat a greater variety of food. Clever, cooperative feeding habits occur in South Carolina and Baja California, where dolphins chase fish onto the shore, then roll up on the beaches, completely out of the water, to grab the fish on the beach. Bottlenose dolphins are sometimes involved with humans, to their mutual advantage, in corralling and catching fish (see p. 227).

Atlantic White-sided Dolphin

Lagenorhynchus acutus

The Atlantic white-sided dolphin was first described scientifically in 1828. Long before this, these dolphins were well known to early fishers and whalers in the North Atlantic. The names for these species of lively dolphins all referred to their habit of jumping clear out of the water as they moved. Norwegians called them "springhval," Germans "springer," and to eastern Canadians, they became "jumpers."

Identifying this dolphin is easy; you just look for the bright yellow patch on the rear flanks. This flashes briefly as the dolphin moves

FIELD NOTES
- Male: 7–9' (2.2–2.7 m); female: 6–9' (1.9–2.7 m); at birth: 3½' (1.1 m)
- Various fish and squid
- Inshore to mainly offshore
- Cold temperate waters of the North Atlantic Ocean
- Insufficiently known

through the water. You may also glimpse the subtle white band on each side below the dorsal fin, which gave this dolphin its common name.

Traveling in groups of up to 100, Atlantic white-sided dolphins are a favorite of whale watchers on tours off Scotland, Ireland, Iceland, Massachusetts, and Newfoundland. The dolphins often accompany whales; they may be feeding on similar prey at times, but for the dolphins, it seems to be more of an opportunity to be sociable. Herds of up to 1,000 dolphins have been seen on occasion in offshore areas.

White-beaked Dolphin

Lagenorhynchus albirostris

The white-beaked dolphin shares most of its North Atlantic range with the Atlantic white-sided dolphin. But it ventures farther north into sub-arctic waters, making it the most northerly occurring of all dolphins. It is the most robust of all the "lag" dolphins (those belonging to the genus *Lagenorhynchus*) with the thickest blubber layer.

In spite of the common name "albirostris," meaning "white beak," the beak, when it can be seen in the wild, is often not

FIELD NOTES
- Male: 8–10' (2.4–3 m); female: 8–10' (2.4–3 m); at birth: 4' (1.2 m)
- Offshore schooling fish, cephalopods, crustaceans
- Inshore to mainly offshore
- Cold temperate to subarctic waters of the North Atlantic Ocean
- Insufficiently known

white, but gray or even black in parts of its range. The best way to distinguish this species from the Atlantic white-sided dolphin is by checking the rear flanks. White-beaked dolphins have a grayish patch; a bright yellowish patch indicates an Atlantic white-sided dolphin.

These dolphins are welcomed on whale-watch tours off Iceland, Norway, Newfoundland and other areas of eastern Canada, Greenland, the Faeroes, and Ireland. They often churn the water, creating, like Dall's porpoise, a "rooster tail" effect.

White-beaked dolphins are powerful swimmers.

Pacific White-sided Dolphin

Lagenorhynchus obliquidens

This exclusively North Pacific dolphin acquired the species name "obliquidens," meaning "slanting tooth," because of its slightly curved teeth. This feature was noted by the fish taxonomist Theodore Nicholas Gill of the Smithsonian Institution, who gave the species its name after examining three skulls that had been collected near San Francisco.

Within their range, Pacific white-sided dolphins are most likely to be confused with common dolphins, as both travel in large groups and have light side patches. However, the Pacific white-sided has a short, thick snout unlike the beak of the common dolphin. It also has a thin, gray stripe along both sides, extending from the head, curving down below the dorsal fin, and ending in a large, white, rear flank patch. Pacific white-sided dolphins often have two-color dorsal fins, dark on the front half and gray or white on the back.

"Lag" dolphins are all acrobatic and sociable. The Pacific white-sided may be the most acrobatic and sociable of them all. They often swim in groups of up to 100, although offshore assemblies of 1,000 to 2,000 are common. Their

FIELD NOTES

- Male: 5½–8' (1.7–2.4 m); female: 5½–8' (1.7–2.4 m); at birth: 3½' (1.1 m)
- Offshore fish and squid
- Mainly offshore but sometimes inshore
- Cool temperate waters of the North Pacific Ocean
- Insufficiently known, killed accidentally in net fishing

sociable natures also extend to other cetaceans, particularly northern right whale dolphins. They often ride the waves— from the bow and wake of boats and from the surf. Intensely curious, they will sometimes inspect boats.

Pacific white-sided dolphins often steal the show on offshore whale-watch trips. They can be seen from British Columbia (Vancouver Island), southeast Alaska, California (particularly southern California and Monterey), Mexico, and Hokkaido, Japan. These dolphins live mainly offshore, but will come closer at certain times of the year, particularly in deep-water areas. A photo-identification study along northeastern Vancouver Island, where they some-times come into inshore waters, has cataloged hundreds of these dolphins.

Acrobatic Pacific white-sided dolphins.

Dusky Dolphin

Lagenorhynchus obscurus

The dusky dolphin was one of two *Lagenorhynchus* dolphins, or "lags," described by John Edward Gray in 1828—the other was the Atlantic white-sided dolphin. Gray saw only a drawing of a dusky dolphin based on a skull and stuffed skin from the Cape of Good Hope, South Africa. He called the species "obscurus" for its dusky coloring.

Resident in the coastal and shelf waters of South Africa, New Zealand, southern Australia and southern South America, duskies in different areas have slightly different patterns on their back, but all have proved to be the same species.

FIELD NOTES

- Male: 7' (2.2 m); female: 6' (1.9 m); at birth: 2' (0.6 m)
- Squid, and fish such as anchovy, bottomfish
- Inshore to continental shelf waters
- Temperate waters of South Africa, New Zealand, southern Australia, and southern South America
- Insufficiently known

At sea, the best way to identify a dusky is to look for the light face and short blunt rostrum with the hint of a beak. The dorsal fin is curved back, pointed, and often has two colors, a variable pattern of dark in the forward area and light in the rear. Look for the two-prong blazes of white, pointing forward on each rear flank.

The duskies are one of the most acrobatic of dolphin species, their slim, light bodies executing extraordinary leaps and somersaults. Once one dolphin starts leaping, others often follow. They seem to enjoy contact with boats and people, and are popular on whale- and dolphin-watching trips from Patagonia, Argentina and Kaikoura, New Zealand.

One of the first-ever dolphin photo-identification studies was carried out on the herds of duskies off Patagonia. Researchers found that the same dolphins stay there throughout the year. In summer, they come together to socialize and rest, but separate into smaller groups to feed.

The highflying, acrobatic breach of a dusky dolphin thrills dolphin watchers off the coast of Kaikoura, New Zealand.

Fraser's Dolphin

Lagenodelphis hosei

This tropical, deep-water dolphin, first described in 1956 and identified in the wild in the 1970s, can now be seen with some regularity on whale-watch trips. In some areas they avoid boats, but in South Africa they ride the bow waves, and have become a welcome feature of whale-watch tours in the Caribbean and the Philippines.

In 1955, cetologist Francis Charles Fraser found a mislabeled skeleton in the British Museum collected 60 years earlier in Sarawak. He placed the specimen somewhere between *Delphinus*, the common dolphins, and *Lagenorhynchus*, the "lag" dolphins, so he invented an intermediate genus: *Lagenodelphis*.

At sea, Fraser's dolphins have a striking color pattern with a dark band stretching from the face to the rear underside and bordered above by a gray or whitish line. The dark band can be very wide and dark in males and its intensity may increase with age. Otherwise, the back is dark brownish gray and the belly is pink or white.

Fraser's dolphins often travel in groups of 100 to 500, and occasionally the group can be as large as 1,000. They sometimes associate or feed with other species of tropical toothed whales and dolphins.

> **FIELD NOTES**
> - Male: 7½–8½' (2.3–2.6 m); female: 7–8½' (2.2–2.6 m); at birth; 3½' (1.1 m)
> - Various fish, squid, shrimp
> - Mainly offshore or deeper waters
> - Subtropical to tropical world waters
> - Insufficiently known

Hourglass Dolphin

Lagenorhynchus cruciger

Hourglass dolphins live in deep waters offshore in the Southern Ocean between Antarctica and the major continents. In the early 1820s, a French expedition off Antarctica watched these dolphins frolic around their ship and named them "cruciger," or cross-bearing. The common name, hourglass dolphin, is derived from the dramatic white, double patches along the flank, which seem to crisscross along the animal's sides. These prominent markings make it fairly easy to identify this robust dolphin.

Not shy, the hourglass dolphin is seen riding the bow or stern waves of fast ships, and swimming parallel to slower vessels. It usually travels alone or in groups of up to 7, but occasionally up to 40 are seen in a group. They may travel with other dolphins and whales. Although these dolphins are encountered more often, their remote habitat and range prevent regular study; there remains much to learn.

> **FIELD NOTES**
> - Male: 5' (1.5 m); female: 6' (1.9 m); at birth: <3½' (<1.1 m)
> - Small deep-water fish and squid
> - Offshore
> - Cold temperate Southern Hemisphere to Antarctic waters
> - Insufficiently known

The hourglass dolphin of the Southern Ocean is easily identified.

Commerson's Dolphin

Cephalorhynchus commersonii

In 1767, as Louis Bougainville's round-the-world expedition approached Tierra del Fuego, Philibert Commerson, a French physician and botanist aboard, noted his delight at seeing striking black-and-white dolphins moving at great speed around the ship. It was not until 1922, however, that these dolphins acquired full recognition and their current scientific name.

Identifying a Commerson's dolphin in the inshore waters of Argentina and Chile, where it is often encountered, is easy. It usually has a distinctive rounded dorsal fin, and dramatic black-and-white markings rather like an orca. A wide, white band between the blowhole and the dorsal fin completely encircles the dolphin. On either side of this white band are the black head and flippers, and the black dorsal fin and rear portion. Around the Kerguelen Islands in the southern Indian Ocean, the dolphin's black-and-white patterns are dark and light gray.

FIELD NOTES

■ Male: 4½–5' (1.4–1.5 m); female: 4½–5' (1.4–1.5 m); at birth: 2½' (0.7 m)
■ Inshore fish, crustaceans, and squid
■ Mainly inshore
■ Cold temperate waters of southeastern South America, Falkland Islands, and Kerguelen Islands
■ Insufficiently known

The shape of the black patch on the underside of a Commerson's dolphin indicates the sex. On a male, it is shaped like a raindrop; on a female it is shaped like a horseshoe.

Commerson's dolphins eagerly ride waves at sea or as they break on shore. They show no fear of boats, often riding the bow and stern waves, but their sometimes erratic swimming behavior makes it difficult to keep track of them. Maneuvering through fast, swirling rapids, they are agile swimmers. Typically, they travel alone or in groups of 2 to 15. Occasionally 100 or more are seen.

Popular with dolphin watchers, Commerson's dolphins can be seen from shore as well as from boats within their limited, remote range. Their movements and behavior are being studied in southern Patagonia, using photo-identification methods.

Easily identified by their striking markings, Commerson's dolphins rest near the Falkland Islands.

174

Hector's Dolphin

Cephalorhynchus hectori

New Zealand's own dolphin, the Hector's dolphin, is one of the smallest cetaceans, at only about 4½ feet (1.4 m) long. It has a complex color pattern of gray, black, and white. Moving in groups of two to eight animals in the waters around New Zealand, they surface frequently to breathe, showing little of their stout bodies. Hector's dolphins differ considerably from bottlenose and other dolphins. Although it is easy to miss them because of their low profile and small size, the best way to identify them is by their characteristic rounded dorsal fin.

Hector's dolphins inhabit muddy river mouths and shallow bays, often swimming among rocks close to the shoreline. They rarely venture more than 5 miles (8 km) from land. They breach, lobtail, spyhop, and engage in most of the dolphin play behavior familiar to watchers.

FIELD NOTES
- Male: 4–4½' (1.2–1.4 m); female: 4½–5' (1.4–1.5 m); at birth: 2'+ (0.6 m+)
- Fish and squid
- Inshore, including estuaries
- Coastal New Zealand waters
- Indeterminate, killed accidentally in gill nets

Mother and calf sighted off Banks Peninsula.

The best place to observe Hector's dolphins is around the Banks Peninsula, in New Zealand's South Island, where regular tours are offered. They can also be viewed from various other ports on the South and North islands, including the popular whale-watch port of Kaikoura. They can even be seen from shore, often near rocks or entering estuaries and rivers, sometimes swimming a little way upstream.

Hector's dolphins can be attracted to boats, particularly slow-moving boats, although they seldom bow ride. Dolphin watchers sometimes find that these dolphins will swim alongside the boat for a time, often swimming in its wake.

In the past, Hector's dolphins were caught for bait, and currently they are trapped accidentally in trawls and especially gill nets. Some parts of their habitat near the Banks Peninsula have been protected, but to save this declining species, more near-shore areas around New Zealand will have to be declared no-go zones for coastal gill nets.

Southern Right Whale Dolphin

Lissodelphis peronii

François Peron, a naturalist aboard *Géographe*, a French expedition to Australia in 1800, was amazed to see these finless dolphins near Tasmania: the first finless dolphins a scientist had ever noted. The common name of this dolphin, and its northern relative, is "right whale" dolphin. (The right whale is many times larger, but also lacks a dorsal fin.)

Compared to the northern right whale dolphin, the southern species has larger white areas on its body. On its low-angle leaps, the white beak and forehead present an

FIELD NOTES
- Male: >7' (>2.2 m); female: <7½' (<2.3 m); at birth: 2½'+ (0.7 m+)
- Squid and a variety of fish (lanternfish)
- Offshore
- Cool temperate to subantarctic waters of the Southern Hemisphere
- Insufficiently known

odd profile. These so-called "mealy-mouthed porpoises" have largely white flippers with either a dark leading or trailing edge, and patches of white on the leading edge of the upper sides of the flukes. The underside is entirely white.

Mainly resident in the cool, deep, temperate waters of the Southern Ocean, these dolphins extend into subantarctic waters. They may be seen on tours near the Falkland Islands, en route to the Antarctic, and off New Zealand, Argentina, Chile, and South Africa.

Northern Right Whale Dolphin

Lissodelphis borealis

This lithe animal with its smooth back, devoid of a dorsal fin, presents a graceful dance as it skims the waves in low-angle leaps. Traveling at speed in groups of up to 200, often with Pacific white-sided dolphins, they look like so many balls bouncing across the water.

Discovered in the mid-nineteenth century, nearly 50 years after its close southern relative (above), the northern species is dark, except sometimes on the tip of its lower jaw, on its chest and the underside of the flukes. Because it has no dorsal fin, it is

FIELD NOTES
- Male: 7–10' (2.2–3 m); female: 6½–7½' (2–2.3 m); at birth: 2½'+ (0.7 m+)
- Variety of fish and squid
- Offshore
- Temperate waters of the North Pacific
- Insufficiently known, commonly caught in net fishing

unlikely to be confused with any of the other cetaceans within its range.

Confined to the cool, deep, temperate waters of the North Pacific, northern right whale dolphins were considered rare until cetacean biologists Stephen Leatherwood and William A. Walker studied them in the mid-1970s. They made numerous sightings, finding as many as 3,000 in one group. Today, these dolphins are more often seen traveling in groups of 5 to 200 on late summer and autumn whale-watch tours to the Monterey submarine canyon. They sometimes bow ride.

A northern right whale dolphin bounces across the surface.

Short-beaked Common Dolphin

Delphinus delphis

One of the most common of all dolphins, this fairly large dolphin is easily distinguished by the hourglass pattern on its side. This makes a dark "V" shape below the dorsal fin, almost like a reflection of the fin. At one time or another, more than 20 species have been proposed and discarded for variations of this geographically widespread species. The case for at least one other species has recently been generally accepted (see below). The description of behavior that follows applies to both forms; specific anatomical

FIELD NOTES

- Male: 5½–7' (1.7–2.2 m); female: 5–6½' (1.5–2 m); at birth: 2½'+ (0.7 m+)
- Small fish and cephalopods
- Mainly offshore
- Temperate and tropical waters, including inland seas (Black Sea)
- Insufficiently known, but taken in many fisheries worldwide

differences are dealt with in the entry below.

Common dolphins travel in groups of 10 to 500 in most areas, with up to 2,000 or more in the eastern tropical Pacific. These herds are so acrobatic and boisterous that the noise of their approach can often be picked up from miles away. Their high-pitched sounds can sometimes be heard as they bow ride.

Although common dolphins are killed in significant numbers in tuna and other fisheries, numbers are thought to be considerable.

Long-beaked Common Dolphin

Delphinus capensis

For decades, it has been known that various stocks of common dolphins look different. In 1994, a new species of dolphin was created by the splitting of common dolphins into short-beaked and long-beaked species. The short-beaked common dolphin kept the original scientific name and the new species took a new name, "capensis." Based on anatomy as well as genetics, the differences between the two species are consistent in at least several parts of the world. However, even within the two species, there is still variation.

The long-beaked common dolphin has a muted color pattern, with less contrast between the dark and the white (or yellow) parts of its body. Subtle differences include a slightly

FIELD NOTES

- Male: 6½–8½' (2–2.6 m); female: 6–7½' (1.9–2.3 m); at birth: 2½'+ (0.7 m+)
- Various small fish and cephalopods
- Coastal
- Temperate and tropical world waters
- Insufficiently known

short-beaked (above) and long-beaked

longer, less chunky body; a less rounded head; and a thicker, dark line between the beak and the flipper. But the most noticeable is that the long-beaked form has a longer beak than the short-beaked form.

Little is known about this dolphin's behavior at sea, apart from what is generally known about common dolphins. However, because of its strong coastal presence, it makes up a sizable proportion of the sightings of common dolphins. **177**

Spinner Dolphin

Stenella longirostris

The spinner dolphin—also known as the rollover, longsnout, long-snouted spinner dolphin, or long-beaked dolphin—is famous for its fantastic spinning leaps, in which a dolphin breaches high out of the water then rolls on its longitudinal axis making up to seven complete turns. Few, if any, dolphins leap as high or as often, and no others, except the clymene dolphin, are known to spin. Spinner dolphins are frequently observed around Hawaii, Mexico, and Japan.

If you can't see them spinning, the best way to distinguish these dolphins from other dolphins within their range is by looking at the long, thin beak, the dark gray stripe from eye to flipper, and the usually well-defined three-toned coloring of the body, ranging from dark on the top, to gray to light on the belly. Also, the dorsal fin, particularly in some populations, stands erect.

FIELD NOTES
- Male: 5½–8' (1.7–2.4 m); female: 5½–7' (1.7–2.2 m); at birth: 2½'+ (0.7 m+)
- Various fish and squid
- Offshore
- Subtropical and tropical world waters
- Insufficiently known, many killed in fishing nets

There are several forms of spinner dolphins. In the eastern tropical Pacific alone, where they have been intensively studied due to their association with yellow-fin tuna, there are three forms, each with slight differences. There are apparently other forms in other parts of their wide range.

At sea, spinner dolphins often approach boats to bow ride and may stay for as long as half an hour—longer than most other dolphins. They travel in groups of 5 to 200, although often up to 1,000 or even more swim in mixed schools with pantropical spotted and other dolphins.

Because of its acrobatic talents, the spinner was one of the first dolphins to be captured for aquariums in the North Pacific, but they have a poor survival record. The main threat to spinners has come from tuna fisheries which have caused the deaths of many hundreds of thousands of spinner dolphins. Although the kills are much fewer today, their numbers have apparently not returned to original population sizes.

The motion of the longitudinal spin can be seen in the twist of the body.

178

Clymene Dolphin

Stenella clymene

Originally described in 1846, the clymene, also known as the short-snouted spinner dolphin, was not considered a valid species until recently. Its external color pattern looked more like the spinner dolphin (see opposite page) with which it had long been confused. In 1975, however, William Perrin and his colleagues found that the skulls of some spinner dolphins from the coast of Texas resembled the original clymene dolphin from the British Museum. In 1981, the clymene dolphin received full species status.

The distribution of the clymene and the spinner overlap in the North Atlantic, although the spinner is found farther south, even into the tropical South Atlantic. Distinguishing spinner and clymene dolphins can be difficult at sea. The clymene is a little more robust than the spinner dolphin. It is also said to have a less pronounced triangular dorsal fin, but this alone isn't sufficient for identification due to the dorsal fin variation among spinner dolphins. Up close, the clymene dolphin has a shorter beak, a dark line which sometimes looks like a mustache on top of the

FIELD NOTES
■ Male: 6–6½' (1.9–2 m);
female: 6–6½' (1.9–2 m);
at birth: 2½' (0.7 m)
■ Small fish and squid
■ Mainly offshore
■ Subtropical and tropical waters
of the Atlantic Ocean including
Gulf of Mexico
■ Insufficiently known

beak, and a darker cape which dips close to the white areas on either side of the dorsal fin.

Normally, the clymene dolphin's leaps are not as high, and its spins not as numerous as the spinner dolphin, but it is the only other dolphin besides the spinner to make longitudinal spins, rather than somersaults. They sometimes bow ride and occasionally approach boats. In feeding areas they associate with other small dolphins such as common and spinner. School size is much less than for the spinner dolphin— usually fewer than 50 animals.

Little yet is known about this dolphin—its distribution, habits, and status. Some are killed for meat by local fishers in the eastern Caribbean.

A dark line on top of the beak helps identify a clymene.

Pantropical Spotted Dolphin

Stenella attenuata

Often called "spotters," or simply spotted dolphins, the pantropical spotted dolphin is a delightful sighting on any marine nature tour. They travel with ships, charging to the bow or stern to ride the waves. They make long, low leaps clear of the water as they swim along, and their breaches are high and frequent, if a little less acrobatic than the spinner dolphin.

There are two recognized species, the pantropical and the Atlantic (see opposite page), but there may be additional species named as more is learned about the taxonomy of these dolphins. In the eastern Pacific alone, there are at least two main forms within the pantropical species, one coastal, the other offshore. The coastal species is larger and more robust, with a thicker beak and more spots.

FIELD NOTES

■ Male: 6½–8½' (2–2.6 m); female: 6–8' (1.9–2.4 m); at birth: 2½'+ (0.7 m+)
■ Small fish and cephalopods
■ Coastal to offshore
■ Mainly tropical, some subtropical and warm temperate world waters
■ Insufficiently known, caught in tuna purse-seine nets, more abundant south of the equator

Even though its spots provide the best clue to identification, this is not always straightforward. At a distance and in some lights, the spots don't always show up, and pantropical populations in the Gulf of Mexico and around Hawaii have few, if any, spots. Bottlenose dolphins and both species of hump-backed dolphins also have some spotting at times, although usually not on the back. Finally, pantropical spotted dolphins are unspotted at birth. As juveniles, they acquire dark spots on their bellies, followed by light spots on the back.

Tuna purse-seine fishing, particularly in the eastern tropical Pacific, has seriously depleted numbers of these dolphins. Pantropical schools are located by plane and ship, and nets are set around them to catch yellowfin and skipjack tuna. Hundreds of thousands of dolphins per year were killed in the 1960s and 1970s until conservation measures began to take effect in the late 1980s. Although not endangered, these dolphins still suffer high losses.

Except for a cookie-cutter shark scar, no spots are obvious on these "spotters" sighted near Hawaii.

Atlantic Spotted Dolphin

Stenella frontalis

S ince the early 1980s, several generations of Atlantic spotted dolphins have been visiting and swimming with dolphin watchers in the clear, shallow waters of the northern Bahamas. The dolphins' movements have been recorded in synchronization with their sounds, allowing for the development of a behavioral ethogram, or catalog of behavioral patterns. Researchers have found generally small schools of 1 to 15 individuals. Larger numbers sometimes join up temporarily. The schools have a fluid structure, with dolphins often joining and splitting into subgroups, as do the well-studied bottlenose dolphins, yet there are also long-term bonds. Like the pantropical spotted dolphin and most other closely related oceanic dolphins, the Atlantic spotted dolphin bow rides, breaches, and plays at every opportunity.

The Atlantic spotted dolphin is found only in the tropical to warm temperate Atlantic. In this area, their range overlaps with the pantropical spotted dolphin and confusion is possible at sea. The main differences between the two, noticeable at close range, are that the Atlantic spotted dolphin is generally more robust, more spotted,

FIELD NOTES
- Male: 6½–7½' (2–2.3 m); female: 6½–7½' (2–2.3 m); at birth: 3'+ (0.9 m+)
- Various fish and cephalopods
- Inshore to offshore
- Tropical, subtropical, and warm temperate waters, of the Atlantic Ocean
- Insufficiently known, caught in Atlantic tuna purse-seine nets

The two Atlantic spotted dolphin juveniles (above) have yet to develop their spots.

and often darker (which helps the spots show up better). The spots develop as the animal matures. As with pantropical spotted dolphins, there are two forms, coastal and offshore, with considerable variation within each group. With an Atlantic spotted dolphin that is not very spotted, typically an offshore form, its robust body is close in overall appearance to that of a bottlenose dolphin.

Beloved of dolphin watchers on the northern fringes of the Caribbean, Atlantic spotted dolphins are still hunted for food in the eastern Caribbean by local fishers. However, cetacean-watch tours from St. Vincent, Grenada, and Dominica are giving visitors and locals the pleasure of watching these animals at sea.

181

Striped Dolphin

Stenella coeruleoalba

The first to appreciate the delicate beauty of dolphins were the ancient Greeks, who painted them in frescoes. Several thousand years later, we can see that their inspiration was undoubtedly the striped dolphin.

These beautiful animals look almost hand-painted. With an upward brush stroke toward the dorsal fin, the light gray flank divides the dark back and the white or pink belly. But the best identifying feature is a thin, dark stripe with a feathery, dark streak below it, extending from the black beak, around the eye patch to the underside of the rear flank. There are also one or two dark bands between the eye and flipper.

At sea, striped dolphins are easy to identify at medium to close distance. Their acrobatic, often aerial behavior ensures that the characteristic stripe will be seen. Like other *Stenella* dolphins, they manage tailspins and somersaults, as well as breaching to great heights, up to 23 feet (7 m), or three times their

FIELD NOTES

- Male: 6–8½' (1.9–2.6 m); female: 6–7' (1.9–2.1 m); at birth: 3½' (1.1 m)
- Various fish, cephalopods, and sometimes crustaceans
- Offshore
- Warm temperate, subtropical, and tropical world waters
- Insufficiently known, accidentally killed in fish nets and hunted in Japan

length. They bow ride in the Atlantic, but there are fewer reports of this behavior in the Pacific or Indian Oceans. Apparently, they are more easily alarmed than other dolphins, and will turn tail and streak away—in the Pacific they are called "streakers." Herds number 100 to 500, but can be up to 3,000. In some areas, such as the Mediterranean, however, group size is fewer than 100.

Like several other dolphins, they sometimes travel above yellowfin tuna, which has led to them being killed, although in fewer numbers than spotted, spinner, and common dolphins. A much more serious threat is Japanese drive fishing, which kills thousands. There are also killings near Sri Lanka, as well as accidental catches during Mediterranean netting operations.

The dark stripe is easily seen on these striped dolphins, as they race along with flying leaps.

Risso's Dolphin

Grampus griseus

The Risso's dolphin is a robust, blunt-headed oceanic dolphin. Larger than any other cetacean that carries the name "dolphin," it is sometimes informally grouped with the "blackfish" because of its size and blunt head.

This dolphin is easy to identify. It is heavily scarred, ghost-like, and looks like no other species, although from a distance the tall, curved-back dorsal fin might be confused with an orca or bottlenose dolphin.

Also called the gray dolphin, or the gray grampus (a literal translation of its scientific name), the Risso's dolphin is gray on the back and sides, and white on the belly. On the head and just in front of the dorsal fin, the coloring is light gray, sometimes white. All appendages (the dorsal fin, flippers, and flukes) tend to be darker than the rest of the body. The body becomes lighter as

FIELD NOTES

- Male: 8½'–12½' (2.6–3.8 m); female: 8½'–12' (2.6–3.7 m); at birth: 4½' (1.4 m)
- Squid and other cephalopods, fish, and crustaceans
- Offshore
- Tropical, subtropical and temperate world waters
- Insufficiently known

As seen here, the torso of the blunt-headed Risso's dolphin is larger than that of the bottlenose dolphin.

the animals age, and older animals can be almost white. Younger animals have few, if any, scars, while mature animals carry a wide range of scratches, blotches, and spots.

Why do these dolphins have so many scars? The Risso's dolphin has between 4 and 14 teeth near the tip of the lower jaw. There are rarely teeth in the upper jaw. Some of the scars come from the teeth of other Risso's dolphins, possibly playing or fighting with each other. However, it is thought that some of the scarring may come from squid bites. Limited stomach studies reveal the dolphin's preference for squid.

Photo-identification studies have been started recently off central California and in the Azores. Risso's dolphins are found in groups of 3 to 50, but sometimes up to 4,000. Compared to other cetaceans, they have been little studied, and not much is known about their status. Many Risso's dolphins have been killed accidentally in gill nets around Sri Lanka.

183

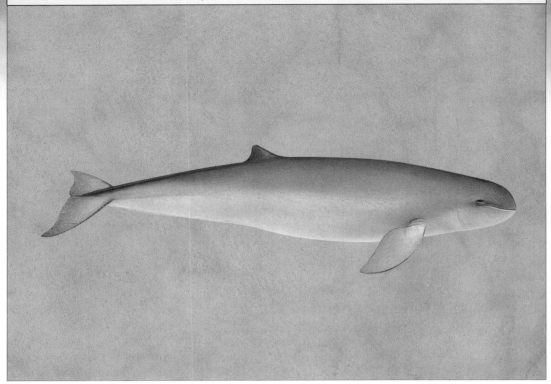

Irrawaddy Dolphin

Orcaella brevirostris

This blunt-nosed, little dolphin with the stubby, almost non-existent dorsal fin makes few waves. Traveling through the inshore waters of southeast Asia, Indonesia, and northern Australia, its low profile is more often missed than recognized. Named after the Irrawaddy River of Myanmar (formerly Burma), it is also sometimes known as the Mahakam River dolphin, after a river in eastern Borneo (Kalimantan) where it lives hundreds of miles upstream.

The Irrawaddy dolphin has many attributes which put it in the family of oceanic dolphins, but some researchers consider it close to the beluga family—it looks a little like a beluga. In the wild, this dolphin can be confused with the finless porpoise which is, however, much smaller and has no dorsal fin. From a distance, they may even sometimes be confused with dugongs.

Irrawaddy dolphins travel in small groups, usually fewer than 6 but sometimes up to 15 individuals. They occasionally make low-angle leaps, lobtail, breach, and spyhop, and they have been known to spit water. However, above-surface behavior is uncommon, and there are no reports of bow riding.

FIELD NOTES

- Male: 7–9' (2.2–2.7 m); female: 7–7½' (2.2–2.3 m); at birth: 2½'+ (0.7 m+)
- Fish, cephalopods, and crustaceans
- Inshore and rivers
- Subtropical and tropical waters of the Indian and western Pacific Oceans
- Insufficiently known

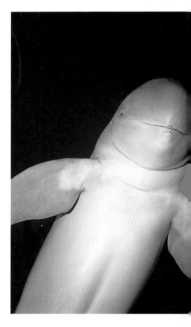

With its blunt, rounded head, indistinct beak, flexible neck, and straight mouth-line, the Irrawaddy dolphin looks a little like a small beluga.

The Irrawaddy's chosen habitat is in tropical rivers, estuaries, and inshore coasts, which means that it comes into frequent conflict with people and industry. It is killed accidentally in shark gill nets in Australia as well as in fish traps and other nets within its range. But more serious threats to the Irrawaddy dolphin arise from the destruction and degradation of its habitat by riverbank development and the construction of dams.

Melon-headed Whale

Peponocephala electra

The melon-headed whale was originally thought to be a "lag" dolphin, similar to the Pacific white-sided dolphin. In 1965, a herd of 500 swam into Suruga Bay, Japan, 250 of which were caught and killed. After studying them, scientists decided this species deserved its own genus.

Melon-headed and pygmy killer whales (see below) are difficult to distinguish at sea. The melon-headed whale, however, has a slender, torpedo-shaped body with shorter, narrower flippers with pointed tips. But the most distinctive difference is the head. The melon-headed whale has a slim, pointed head,

FIELD NOTES
- Male: 7–9' (2.2–2.7 m); female: 7½–9' (2.3–2.7 m); at birth: 2'+ (0.6 m+)
- Various small fish and squid
- Offshore
- Tropical and some subtropical world waters
- Insufficiently known, killed accidentally through gill nets and other fishing methods

while the pygmy killer whale has a rounded head.

Melon-headed whales generally travel in large, tightly packed herds of 100 to 500, with occasional herds of 1,500 to 2,000. They may bow ride and spyhop, but usually steer clear of boats. They have a reputation for being fierce; when two groups were captured for marine aquariums in the Philippines and Hawaii, they attacked their handlers. The best places to find melon-headed whales are off Cebu Island in the Philippines, the east coast of Australia, Hawaii, and in the eastern Caribbean, especially around Dominica.

Pygmy Killer Whale

Feresa attenuata

The pygmy killer whale is as little known as the melon-headed whale with which it shares almost the same habitat and range. It also has a similar appearance. It has been reported to herd and attack dolphins in the South Atlantic and the tropical Pacific. Like melon-headed whales, pygmy killer whales have been captured several times for marine aquariums and have proved aggressive.

Compared to the melon-headed whale, the pygmy killer whale has a more rounded head and flipper tips, and a darker cape. Skull comparisons of stranded specimens also show fewer teeth. In general, they travel in much

FIELD NOTES
- Male: 6½'–9½' (2–2.9 m); female: 7–8' (2.2–2.4 m); at birth: 1½'+ (0.5 m+)
- Various smaller fish and squid
- Mainly offshore
- Tropical and subtropical world waters
- Insufficiently known

Pygmy killer whale, off Florida's Atlantic coast.

smaller groups—fewer than 50 is typical. They are commonly encountered near Dominica, around St. Vincent, Hawaii, and parts of southern Japan. They occasionally bow ride and breach, but aerial behavior is otherwise rare. Most of the time, they tend to avoid boats. A glimpse of these whales is an unusual and welcome sight.

185

False Killer Whale

Pseudorca crassidens

The false killer whale, sometimes also called "pseudorca," is a warm-water resident of the world's oceans. Like orcas and pilot whales, it has a complex social nature. Smaller than pilot whales and orcas, the false killer is much larger than any other "dolphin." All three are among those sometimes called "blackfish."

The false killer whale has a long, slender head and slim body. The dorsal fin looks like a young orca's—prominent and curved back and not wide at the base like a pilot whale's. Compared to two other blackfish, the pygmy killer whale and melon-headed whale with which it is most often confused, the false killer whale is much longer and larger. The head is all black (no white lips, as found in some other blackfish). The flippers are unique, if you can manage to glimpse them as the animal swims along or bow rides: there is a prominent hump that looks like a bend or elbow halfway along the leading edge of each flipper.

FIELD NOTES
- Male: 12–19½' (3.7–5.9 m); female: 11½–16½' (3.5–5 m); at birth: 5½' (1.7 m)
- Various fish, cephalopods, rarely dolphins
- Mainly offshore
- Tropical, subtropical, and sometimes temperate world waters
- Insufficiently known

False killers are fast, acrobatic swimmers—acting more like playful, inquisitive dolphins than pilot whales or orcas. They often breach, sometimes landing on their sides or backs with a great splash. They travel in pods of 10 to 50, although several hundred are sometimes seen in a superpod, or grouping of more than one pod. Both sexes and all ages travel together. False killer whales are often involved in mass strandings; the largest recorded was of more than 800 animals.

False killers are hunted off China, Japan, and the Caribbean, and they are also sometimes killed accidentally in fish nets, including tuna purse-seines and pelagic gill nets. They have a reputation for stealing fish off lines and from nets. They have also been reported to attack dolphins escaping from tuna nets and there is one case of them attacking and killing a humpback whale calf near Hawaii. Their usual diet, however, is fish and cephalopods.

The slim body of a false killer whale is seen when it breaches.

Orca

Orcinus orca

The largest of all dolphins, the orca is found in all seas, from the equator to the polar ice. It is among one of the most widely distributed animals on Earth. Until quite recently it had the reputation of a fierce killer. Its other name, "killer whale," comes from eighteenth-century whalers who saw orcas feeding on other whales and dolphins. As top predator in the sea, the orca's diet extends to several hundred known species—a more diverse and extensive diet than that of any other whale or dolphin. There is no known case of a wild orca ever killing a human.

Its white eye patch, gray saddle patch, and black back make the orca easy to identify at sea. The under-side is entirely white and the flippers are all black. The tall dorsal fin of the male, unique among cetaceans, can be recognized even at a great distance. In a mature male, it can be up to 6 feet (1.9 m) high; at half the male's height,

FIELD NOTES

- Male: 17–29½' (5.1–9 m); female: 15–25½' (4.6–7.7 m); at birth: 6'+ (1.9 m+)
- Numerous fish, squid, and marine mammal species, including blue whales; even gulls, penguins, turtles
- Inshore and offshore
- Equatorial waters to polar ice
- Insufficiently known, hunted and captured in recent years off Japan, Iceland, Antarctica, and other areas

A mother orca, on the right, travels with her two offspring, a juvenile and a full-grown male.

even the female dorsal fin is tall. The flippers are broad and rounded and more than 6 feet (1.9 m) long in mature males.

A 25-year study off British Columbia and Washington State, has found that orcas stay in long-term social groups, or pods, for life, an average of 29 years for males and 50 years for females. Residents live in close family pods of 7 to 50 and subsist on fish; transient pods of 1 to 7 feed on marine mammals. The two groups do not mix.

Orcas can be reliably seen on tours around Vancouver Island, off Antarctica, Norway, and Iceland, and occasionally in many other areas. They are very curious and will approach close to boats to inspect them. Traveling in close-knit groups, they make a great display of activity, spouting loudly, spyhopping, breaching, and lobtailing. Youngsters will sometimes ride the bow or wake. Orcas have been hunted in the past few decades, some of them taken alive as part of the world aquarium trade.

187

Short-finned Pilot Whale

Globicephala macrorhynchus

Short-finned pilot whales range through offshore, warm temperate and tropical seas, making deep dives mainly to feed on squid. Pilot whales are so intensely social that they are almost never seen alone.

Identifying short-finned pilot whales is fairly easy. They sometimes travel with dolphins, such as bottlenose dolphins, but they are much larger and their black bodies are darker—most dolphins tend to be gray. The short-finned pilot whale's dorsal fin is broad based and set far forward on the back. Its head is rounded and bulbous. The only animal likely to be confused with this whale is its close relative, the long-finned pilot whale (see opposite page).

Short-finned pilot whales resident in the warm waters off Tenerife in the Canary Islands have been studied, with 445 individuals photo-identified as part of

FIELD NOTES

■ Male: 20–22' (6–6.7 m); female: 17–18' (5.1–5.5 m); at birth: 4½'+ (1.4 m+)

■ Squid

■ Offshore

■ Warm temperate and tropical world waters

■ Insufficiently known, exploited to some extent, but not extensively

long-term behavioral research. Here, scientists have found that the whales prefer water depths of about 3,300 feet (1,000 m). Typical group size is 10 to 30, but some pods are as large as 60. Like orcas, pilot whales live a long time, with females known to survive up to 65 years.

Short-finned pilot whales rarely breach, but sometimes spyhop. They can be seen on whale-watch trips in the Bahamas; in the eastern Caribbean; in the Azores; in Kochi Prefecture, off Japan; in the Tañon Strait, the Philippines; and around Hawaii.

Although not hunted to the same extent as long-finned pilot whales, several hundred short-finned pilot whales are killed every year off Japan, as well as a few in the eastern Caribbean by local fishers. In Japanese waters, there are two forms of the species.

Both pilot whale species have some variation and separation of populations. In future, this may lead to scientists isolating another species.

188 *Short-finned pilot whales with a yearling calf, Azores Islands.*

Long-finned Pilot Whale

Globicephala melas

B oth species of pilot whale are sometimes called "potheads," a name that was given, as many whale names were, by the hunters who first encountered them. The name is based on the resemblance of the pilot whale's head to a black iron cauldron, or pot. The long-finned pilot whale is still hunted persistently in the North Atlantic Ocean (see p. 47).

Like the short-finned pilot whale, the long-finned is a family animal, traveling in groups of 10 to 50, and sometimes up to 100 or more. There are reports of thousands seen together in great superpods. Any sighting of a group of black, medium-sized whales with rounded heads and very wide, thick, curved-back dorsal fins, is bound to be of pilot whales. Young long-finned pilot whales will sometimes breach, but this is rarely observed in an adult.

In a few temperate waters of the world, the distribution of short-finned and long-finned pilot whales overlaps, making it difficult to tell the two apart. Externally they look alike, except that the long-finned has longer flippers (pectoral

FIELD NOTES
- Male: 13–25' (4–7.6 m); female: 10–18½' (3–5.6 m); at birth: 6' (1.9 m)
- Squid
- Offshore
- Cold temperate to subpolar North Atlantic and Southern Ocean
- Insufficiently known, extinct in North Pacific, still hunted in North Atlantic

long flipper

short flipper

fins) and a few more teeth. It is difficult, however, to see flippers, much less teeth, at sea. At times, positive identification may be impossible. It's no wonder that mariners usually write "pilot whales" in their logbooks, refusing to try to distinguish between the two. But in most areas, distribution is quite enough to determine the species.

Long-finned pilot whales inhabit cold temperate to subpolar waters in both the Northern and Southern Hemisphere, except in the North Pacific. The species lived in the North Pacific off Japan until at least the tenth century, but has since completely disappeared.

The long-finned pilot whale often strands on beaches and is possibly subject to more mass strandings than any other cetacean. Frequent strandings occur on Cape Cod, in southern Australia, New Zealand, and southern South America. Because of their social ties, when one pilot whale strands, the others remain with it. **189**

Narwhal

Monodon monoceros

The tusk of the narwhal may well be the source of the unicorn myths. The tusk is actually a tooth. Narwhals have two teeth, both in the upper jaw. In females, they rarely erupt. In males, the left tooth erupts, penetrating the upper beak and spiraling out up to 10 feet (3 m). The tusk is primarily for display and is used to compete for females (the largest tusked male may get the female). The scratches on the head of many males may result from comparing tusks, and perhaps a little sparring. The sound of two tusks hitting is like the "clack" from musical sticks. The tusk is mostly hollow and an estimated one in three tusks is broken.

FIELD NOTES

- Male: 13–20½' (4– 6.2 m), not including tusk; female: 11– 16½' (3.4 – 5 m); at birth: 5'+ (1.5 m+)
- Wide variety of fish including herring, cod, halibut, salmon; cephalopods, crustaceans
- Mainly offshore
- Arctic Ocean and adjacent bays and straits (except western Canada to eastern Russia)
- Insufficiently known, some taken by native hunters

the erupted left tooth of a male

The narwhal, with the beluga, is a member of the family of "white whales" (Monodontidae) which are considered whales by some and dolphins by others. They have many characteristics of larger dolphins, but also qualify as a separate family of toothed whales.

Tusked narwhals are unmistakable. Only the females and calves, which normally live in separate groups from the males, may be confused with belugas. Narwhals and belugas have a similar head with a bulbous forehead and the hint of a beak. Narwhals sometimes turn whiter with age, but most are considerably darker and splotchy compared to the uniformly light color of the beluga. Narwhals have only a slight hump instead of a dorsal fin, whereas belugas have a dorsal ridge that shows up as a series of dark bumps. Finally, narwhals have distinctive flukes with convex trailing edges that make the flukes appear to be put on backward.

Narwhals have been hunted for centuries by European and Inuit peoples, and today they are still hunted in northern Canada and Greenland.

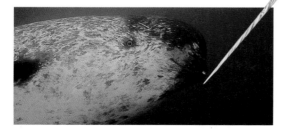

A narwhal's splotchy skin distinguishes it from a beluga.

Beluga

Delphinapterus leucas

The beluga, sometimes called the white whale, is one of the most vocal of all cetaceans. The sounds can often be heard clearly above water or through the hull of a boat.

With their all-white, sometimes yellowish, body, belugas are distinctive. The calves are dark to brownish gray at birth and whiten as they age, reaching pure white between the ages of 5 and 12.

As an arctic animal, the beluga is robust and blubbery. The body is wrinkled and flexible to an extent found in few other cetaceans. The head with its rounded forehead is small compared to the body. The forehead changes shape and the lips can appear rounded as they vocalize. Belugas have a visible neck and can move their head from side to side, giving them the ability to look behind.

At sea, belugas can be difficult to find. With a small dorsal ridge and no fin, belugas have a low profile, and they rarely leap. However, they swim slowly, surface often, and their white bodies contrast with a dark sea. They spyhop often and show curiosity toward boats. In rivers, such as the Churchill in Manitoba, or the St. Lawrence-Saguenay, finding them is much easier.

FIELD NOTES

- Male: 12–18' (3.7–5.5 m); female: 10–13½' (3–4.1 m); at birth: 5'+ (1.5 m+)
- Large variety of fish, various crustaceans and other bottom-living invertebrates
- Inshore including estuaries and rivers in summer, sometimes offshore
- Arctic and subarctic waters (including St. Lawrence River)
- Insufficiently known, population reduced by hunting

An inquisitive beluga shows its flexible neck and blubbery body.

Belugas travel in groups of 5 to 20, although more than a thousand may congregate around the estuaries and in rivers to feed in the summer. They have been known to swim hundreds of miles up rivers in Russia, Canada, and northern Europe. They have little fear of shallow waters, and if stranded are often able to wait and refloat on the next tide.

Belugas have been hunted by Russian, European, and native people for many centuries. A large number of belugas are still taken for food, and numbers are down in some areas. Greater concerns, however, come from the effects of oil and gas activities and chemical pollution.

Dall's Porpoise

Phocoenoides dalli

The hyperactive Dall's porpoise acts more like an excited dolphin than a shy, retiring porpoise. Resident in the cold temperate North Pacific, Dall's porpoises will race around whale-watching boats, even riding the bow. Traveling just beneath the surface at estimated speeds of up to 35 miles (56 km) per hour, they push up a distinctive spray that looks like a "rooster tail." They don't leap out of the water, but the excitement they create on a whale-watch tour can last for hours.

FIELD NOTES

- Male: 6–8' (1.9–2.4 m); female: 6–7' (1.9–2.2 m); at birth: 2½' (0.7 m)
- Hake, herring, mackerel, capelin and other small fish; squid
- Inshore and offshore
- Warm temperate to subarctic waters of the North Pacific Ocean
- Not well known, many killed in western Pacific hunts and in fishing nets

Most Dall's porpoises are black with a white belly and lower flanks, and with fringes of white on the tail and dorsal fin. There are several forms of the species and the various color patterns on the dorsal fin and body may represent subspecies or races. Dall's porpoises are two to three times the bulk of other porpoise species. They were first noted by American zoologist William H. Dall as large, "porpoise-like" animals. His sightings were made off the coast of Alaska in the 1870s.

Harbor Porpoise

Phocoena phocoena

The harbor porpoise is the most commonly seen and studied member of its family, even though it is generally wary of boats and little of its body shows when it surfaces. Once glimpsed, it can be recognized by its low dorsal fin and absence of a beak. The upper body is gray to black, including the small flippers and flukes. The white on the belly and flank turns to gray as it extends high up the sides.

The blow of the harbor porpoise is rarely seen but can be heard. Whale watchers unimpressed at first with the idea of seeing porpoises, come to enjoy the familiar "pop" of its spouting.

FIELD NOTES

- Male: 6' (1.9 m); female: 6' (1.9 m); at birth: 2½' (0.7 m)
- Variety of inshore and offshore fish, herring, mackerel, and anchovy
- Inshore to offshore
- Coastal waters of temperate to subarctic North Atlantic and North Pacific
- Still exist in large numbers

Found mainly in cool, coastal waters throughout the Northern Hemisphere, the porpoise travels in groups of two to five, surfacing about eight times a minute with a slow, forward-rolling motion. When feeding, it might surface to breathe about four times every 10 to 20 seconds before diving for 2 to 6 minutes.

It is the shortest lived cetacean, rarely surviving past the age of 12 years. Even this short life is threatened by humans. It gets caught in fishing nets, is still hunted in a few areas, and has suffered habitat loss near urban areas and shipping lanes.

Harbor porpoises tend to travel in small groups.

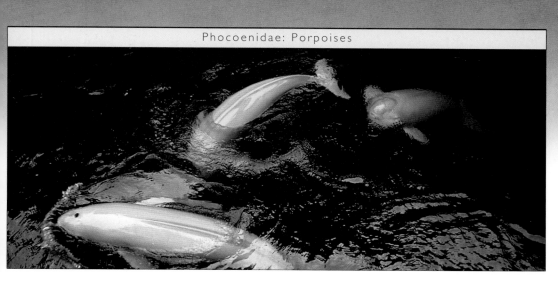

Finless Porpoise
Neophocaena phocaenoides

T his light–colored Asian porpoise is distinguished by its lack of a dorsal fin. Instead, it has a low ridge studded with small, wart–like tubercles along its back. It is one of the smallest cetaceans, and with its bulbous forehead, it looks a little like a tiny beluga.

Finless porpoises often travel in groups of up to 10 individuals, and are found feeding in rivers, estuaries, and mangroves where fresh and salt water mingle. They also travel upriver and out to sea, but most are found no more than 3 miles (5 km) from shore. Like the closely related Dall's and harbor porpoises, the finless porpoise is an

FIELD NOTES
■ Male: 5–6' (1.5–1.9 m); female: 4½–5½' (1.4–1.7 m); at birth: 2' (0.6 m)
■ Small fish; prawns; squid; cuttlefish
■ Coastal waters, estuaries, and rivers
■ Tropical and warm temperate waters from Japan to the Persian Gulf
■ Not well known, some populations reduced by habitat destruction

active animal, and can often be seen darting from side to side as part of its routine. Usually only 1 or 2 are seen at a time, but groups of up to 50 have been reported in feeding areas.

They sometimes spyhop, and at such times the eyes may be visible—about half of all finless porpoises have bright pink eyes.

Living near large population centers in Asia, finless porpoises have suffered considerable habitat destruction from shipping traffic, dams, and pollution. It must be expected that the tremendous human population growth in Asia will continue to have an impact on this species for decades to come.

Vaquita
Phocoena sinus

V aquitas are found only in shallow lagoons in and around the Colorado River delta in the northern Gulf of California, Mexico. They are locally called Gulf of California porpoise, or cochito. Studies of the animal often involve long periods of searching with few sightings. Like some other porpoise species, vaquitas may well avoid boats.

One of the smallest cetacean species, the inconspicuous vaquita grows to only 5 feet (1.5 m). Its distinctive triangular fin is large compared to the rest of the body, and at first sight might seem shark–like. Most "encounters" have been with dead animals

FIELD NOTES
■ Male: 4½' (1.4 m); female: 5' (1.5 m); at birth: 2½' (0.7 m)
■ Grunts, gulf croakers, and other fish; squid
■ Inshore
■ Warm temperate waters in the northern Gulf of California
■ Endangered, may be too few left for the species to recover

A vaquita retrieved from a gill net.

accidentally caught in gill nets. Despite bans, illegal gill netting has persisted for many years. In June 1993, the Mexican government established the Upper Gulf of California Biosphere Reserve to protect the vaquita and its habitat. It remains to be seen whether there are enough left to save the species. **193**

*The greatest resource of the ocean is not
material but the boundless spring of inspiration
and well-being we gain from her.*

JACQUES COUSTEAU (1910–97), French diver and photographer

CHAPTER EIGHT

WATCHING WHALES, DOLPHINS, *and* PORPOISES

USING *the* GUIDE *to* WATCHING CETACEANS

Opportunities to watch and learn about cetaceans are offered in key areas of the world, from the Arctic to Antarctica.

The following pages feature 30 special areas, divided into 7 regions, all of which are exciting and reliable places to watch whales, dolphins, and porpoises. The species you are likely to encounter, and the behavior you may observe, are complemented with information, as well as tips, on how best to watch these magnificent creatures. This guide will give you an insight into what to expect in each area, alert you to the unexpected, and help you select a whale-watch trip according to your interests.

*The **illustrated banding** indicates that the spread is about a special area.*

Watching Whales, Dolphins, and Porpoises

Vivid photos *give a fresh perspective on whales, dolphins, or porpoises that may be encountered in each area, or of unique features that may be of special interest to the visiting whale watcher.*

The **text** *describes the ambience of an area, or sometimes of a particular spot, and how to get there. It lists some of the species likely to be found, and why the animals are there, and explains some of the behavior you may observe. It also points out unusual cetaceans or other attractions you may want to include in your trip.*

Hawaii
West Coast of North America

Every Christmastime, humpback whales return to sing, fight, mate, and raise their calves in the tropical blue waters off Hawaii. Humpbacks make their winter homes near these islands, which are checkered with tropical forest, pineapple and coffee farms, and topped with volcanoes. From hotels on the west side of Maui, it is possible to watch the humpbacks as they arrive; they make occasional approaches opposite Lahaina's Front Street, so be prepared.

Migrating from Alaska, the female humpbacks often arrive first, some heavy with calf, others having just given birth. The males arrive later in search of a partner. In the late 1970s, biologists in Maui first witnessed humpback whales fighting. The fighting seems to follow extended bouts of singing among the males who apparently compete to escort and mate with females. Many of the females start these fights or get involved in them by emitting streams of warning bubbles toward ambitious males. Blood can sometimes be seen on the tail or back of a whale or in the water, but the injuries are usually only superficial.

A humpback nearly clear of the water as it breaches (above) close to shore. A spinner dolphin (left) executes a breathtakingly high, spinning leap.

The shallow, warm sea between the islands of Maui, Molokai, and all around Lanai is the prime area in Hawaii for humpbacks and the core area of the Hawaiian Islands National Marine Sanctuary, which was set up mainly to protect these whales. The best access is through boat tours from Lahaina or Kihei on Maui's west side. Maui is a short plane flight from the international airport at Honolulu on the nearby island of Oahu.

The humpback tours usually include sightings of bottlenose dolphins, which often accompany

208

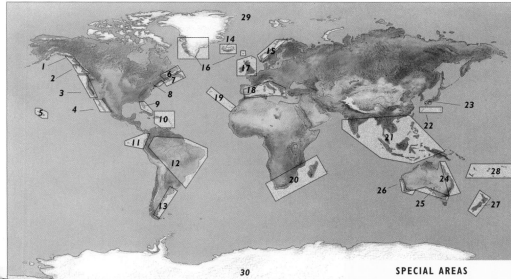

Hawaii

Close behind the powerful tail of a short-finned pilot whale (above). The rugged, volcanic coastline of Kauai (left).

acks. As in other dolphins will ride a wave, but here they se of a whale as well. the humpbacks ated with the panionship, and ve seen humpbacks s at the dolphins, urage them and y. Beneath the beside your boat, e dolphins— wear polarizing the glare—or even whales abbles from their blowholes ough the water. her main Hawaiian islands offer watching and a chance to see a pical cetaceans. Although focus on the humpbacks, will see include bottlenose s, false killer and even the From the big ner high leaps n through hroughout er cetacean anging ve from

SPECIAL FEATURES

From the main street of Lahaina, signs of whale-watching culture date from the 1970s. In front of the Pioneer Hotel, once home to whaling captains waiting for South Seas postings, about 20 whale-watch boats are berthed, each with its signboard displayed. You can see glass-bottom boats, Chinese junks that look like pirate ships, motor cruisers, Zodiac inflatables, schooners, and trimarans. This is a rare opportunity to go window-shopping for the best whale-watch tour. Key features to look for are an on-board naturalist or nature guide, and a boat equipped with hydrophones, through which you can hear haunting humpback songs.

TRAVELER'S NOTES

When to visit *Late Dec–Apr for humpback whales; largest concentrations in Feb and Mar. Year-round: spinner, bottlenose, and pantropical spotted dolphins, false killer whales, short-finned pilot whales, and many others*

Weather *Warm to hot; sometimes strong Kona winds bring rain and high seas*

Types of tours *Half- and full-day tours; some extended multi-day*

expeditions; inflatables, sailboats, and large whale-watch boats

Tours available *Maui: Lahaina, Kihei, Maalaea; Hawaii (big island): Keauhou, Kailua-Kona, Kohala Coast, Honokohau; Oahu: Honolulu, Kaneohe; Kauai: Hanalei*

Information *Hawaii Visitors Bureau, 2270 Kalakaua Avenue, Suite 801, Honolulu, HI 96813; Ph. 1 808 924 0266*

209

Traveler's Notes: These provide brief details on the best time to see cetaceans, the weather, and possible sea conditions. They also list what types of tours are available, points of departure, and contact details for bookings or further information.

WEST COAST
of NORTH AMERICA

Home of the earliest whale-watch tours, the West Coast offers excellent opportunities to sight a wide diversity of cetaceans.

The world's first commercial whale watching occurred in southern California in early 1955, when Chuck Chamberlin, a fisherman from San Diego, put out a hand-made sign that said "See the Whales—$1." The whales were gray whales that migrated along the west coast of North America every winter. The trips proved a steady seller and, four years later, Raymond M. Gilmore, a marine scientist, began the first trips to be led by a naturalist. By the late 1960s, Gilmore and others were leading trips to Baja California to encounter the gray whales mating and raising their calves in Scammon's Lagoon.

At the same time, more and more people began to gather at coastal California lookouts, especially during weekends in the winter, to

A juvenile gray whale near the Channel Islands, off California.

look for spouting gray whales. The whale-watching craze spread north and south along the coast, following the gray whales' migration path.

In the early 1970s, as whaling finally came to a halt around North America and the whales started returning to coastal waters, more than 20 fishing and other communities became involved in the whale-watching business. From California, north to Washington State, and south to Baja California, whale watching became a way for

those who fished or worked in the travel industry to earn some money through winter.

Whale watching grew steadily on the West Coast and soon spread to Hawaii. At first, humpback tours were conducted from Lahaina on Maui, but today, tours leave from all the main islands and encounter a variety of tropical dolphins as well as the humpbacks.

Alaska joined the ranks, with humpback whales again the main attraction. In British Columbia, the orcas became the first toothed whales to be commercially watched, and trips to see them soon rivaled the popularity of tours to spot some of the large baleen whales.

In the late 1970s, seabird and other tours to the waters off San Francisco and Monterey, California, began turning up some amazing finds from August to October: blue and humpback whales, and a wide assortment of other species. As the whale-watching business flourished, these important areas for whales, birds, and fish soon became candidates to be declared marine-protected

A humpback close to the shore of Lahaina, Maui, Hawaii.

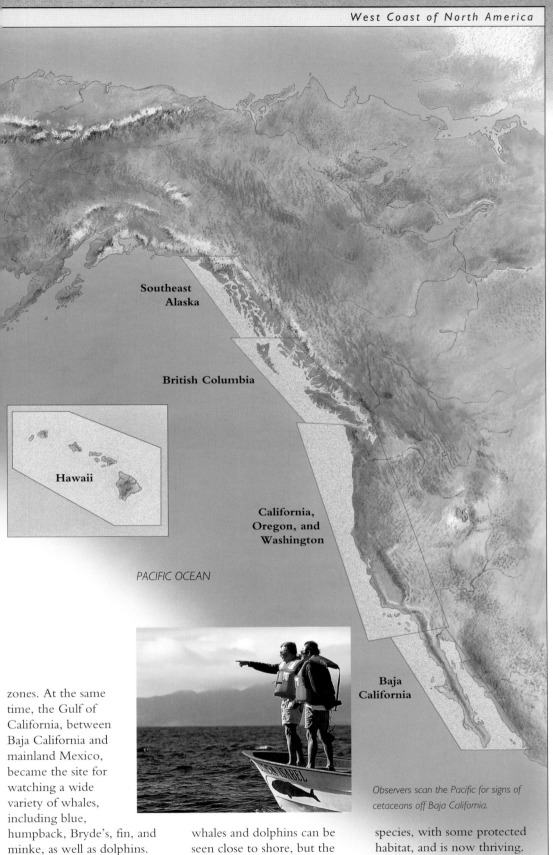

Southeast
Alaska

British Columbia

Hawaii

California,
Oregon, and
Washington

PACIFIC OCEAN

Baja
California

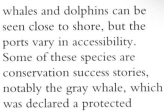

Observers scan the Pacific for signs of cetaceans off Baja California.

zones. At the same time, the Gulf of California, between Baja California and mainland Mexico, became the site for watching a wide variety of whales, including blue, humpback, Bryde's, fin, and minke, as well as dolphins.

Opportunities for whale watching along the Pacific coast are among the most diverse and of the highest quality in the world. Most

whales and dolphins can be seen close to shore, but the ports vary in accessibility. Some of these species are conservation success stories, notably the gray whale, which was declared a protected

species, with some protected habitat, and is now thriving.

Fortunately, the pressures of human population are not as severe here as on the United States east coast, the St. Lawrence, and in Europe.

199

Southeast Alaska

West Coast of North America

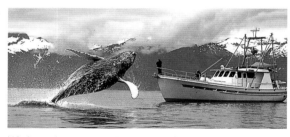

Whale watchers are treated to the sight of a humpback breaching, its flippers outstretched as if taking flight (above). A minke whale (below).

The grandeur of the humpback whale fits the majesty of southeast Alaska, where there is the feeling of being in the presence of nature without bounds. The visitor can sense what the world must have been like before humans arrived on the scene: deep fiords filled with salmon, small schooling fish, and plankton; walls of mountains with tall, dense stands of conifers; glaciers moving at their own slow pace; lush, untouched islands; open sea; and fins and flukes wherever you look.

Southeast Alaska is where "bubblenet feeding" (see p. 66) by humpbacks was first seen and studied. The humpbacks come to Alaska each year from Mexico and Hawaii, and use this technique to round up small schooling fish.

Besides humpbacks, various other whales and dolphins are seen. Orcas patrol the straits and island passages in pods of 10 or more whales. They are here to feed on salmon, although other orcas, the so-called "transients," come to feed on porpoises, seals, and other marine mammals, which sometimes includes whales. Judging from the orca teeth marks on many humpback whale tails, they try, but are not often successful. Minke whales, feeding on plankton and small schooling fish, can also be found all along the southeast Alaskan coast, as well as Dall's and harbor porpoises, Pacific white-sided dolphins and, less frequently, fin whales.

Whale-watching vessels in southeast Alaska range from giant cruise ships to tiny one-person kayaks. Some tours use medium-size motor boats in and around Glacier Bay and Gustavus, the main port for whale-watch boats in southeast Alaska. These tours are day trips, but the cruise ships and kayaks make multi-day tours.

Alaska is the world's number three cruise ship destination; only the Caribbean and Europe

Humpbacks, after a cooperative "bubblenet feed" (above).
Watchers on a cruise ship (left) in Glacier Bay.

attract greater numbers. Summer cruise ships run from California, Seattle, and Vancouver through the Inside Passage to various ports in southeast Alaska. Twenty years ago, cruise ships did not stop for whales. If whales were seen, it was in passing and a matter of luck. Cruise ships then began to advertise their tours using pictures of orca pods and humpback whales as they cavorted against the backdrop of tall conifers, mountains, and glaciers. Customers now demand that ships stop for whales and to be guaranteed sightings.

Today, many cruise ships take a few hours or even a half-day excursion to spend time with the whales, usually in Icy Strait, near Gustavus, and Glacier Bay. Cruise ships are too large and unmaneuverable to approach close to wildlife, but for many people, this is their only chance to see whales. Some cruise ships launch inflatable boats for close-up observation, but the best whale watching normally occurs from medium-size tour boats, such as dedicated whale-watch boats, sailboats, and converted fishing boats.

SPECIAL FEATURES

Wildlife is a big part of all tours to southeast Alaska. Besides the high-profile humpback whales and orcas, you can see harbor seals and Steller sea lions on remote rocky islets. Southeast Alaska has the highest density of black bears in the world. To see brown bears and eagles feeding on spawning salmon in the rivers, the Pack Creek Cooperative Management Area/Stan Price State Wildlife Sanctuary has camping and guided trips in July and August. Access is gained from the west side of Seymour Canal and Admiralty Island, 28 miles (45 km) by air or boat south of Juneau. Another popular site, farther south, is Anan Creek Bear Observatory, on the mainland near Wrangell Island.

TRAVELER'S NOTES

When to visit June–early Sept humpbacks, minke whales, orcas, Pacific white-sided dolphins, Dall's and harbor porpoises. Bubblenetting humpbacks best in June–early July

Weather June–Sept cool to cold on water and subject to extreme changes, including fog and rain; Aug prime month for weather, seas, and whales

Types of tours Multi-day cruises on small and large cruise ships and kayak expeditions; day trips on fishing, sail, inflatable, and whale-watch boats

Tours available Gustavus, Pt. Adolphus, Glacier Bay, Ketchikan, Juneau, Petersburg, Elfin Cove, Wrangell Island, and Sitka; also long-range cruise ships from Seattle, Vancouver, Prince Rupert, and San Francisco

Information Alaska State Division of Tourism, E-28, Juneau, AK 99801 Ph. 1 907 465 2010

British Columbia and Puget Sound

West Coast of North America

Along the British Columbia and Washington coast, the towering dorsal fins of orcas cut through the mist. They have become emblems of the Northwest—visible in everything from Kwakiutl and Haida native totem poles to road signs and tourist brochures. At sea, the tall angularity of the fins matches the tall, snowcapped peaks behind. But more often, it is the sound of orcas spouting, like shotgun fire on a still morning, which alerts you to their approach.

You may see orcas almost anywhere through the Inside Passage or outer coast—from land, ferry, cruise ship, or whale-watch tour boat. The best chance, however, is in two prime spots—western Johnstone Strait-Blackfish Sound (the better spot) and southern Vancouver Island.

Many of the boat tours for western Johnstone Strait-Blackfish Sound depart from Telegraph Cove, a town on northern Vancouver Island. Telegraph Cove, where the main street is a boardwalk that winds along the water, is a half-day drive north from Victoria, or you can fly in

Pacific white-sided dolphins delight watchers with acrobatics off the British Columbia coast.

to Port Hardy from Vancouver and drive to "the cove." Tours will escort you on half-day or full-day trips to meet the orcas.

A typical whale watch usually starts with bald eagles, harbor seals, and sometimes sea lions. Orcas and other cetacean species, including Dall's porpoises, harbor porpoises, minke whales, and Pacific white-sided dolphins, are often seen. More than 10 pods of orcas regularly patrol Johnstone Strait. Although they come mainly to feed on salmon, a day in the life of an orca pod includes periods of rest, play, and socializing. At such times, they often interact with whale-watch boats, playing up for the cameras. A little farther out to sea, off the north and southwest coast of Vancouver Island, humpback whales are sometimes found, but these areas are less accessible to marine tourism. Some tours will end the day with a salmon barbecue back at the port.

The other good spot for orcas is off southern Vancouver Island. Resident to the waters of this

A photographer's paradise: an orca (above) breaches with the tree-clad coast of San Juan Island as a spectacular backdrop; and (right) a pod of orcas in formation with a bull in the foreground.

area and northern Washington State, are 3 pods, comprising about 100 whales. The best boat access is from either Victoria, or Friday Harbor on San Juan Island in Washington. You can reach San Juan Island by car ferry from Anacortes, Washington. In Friday Harbor, don't miss the Whale Museum on First Street. This is dedicated to orcas and other marine mammals in the area and has both research and educational programs. On the west side of San Juan Island there are good land-based lookouts for orcas.

Gray whales migrate along the length of the British Columbia coast. The best place to see them is by boat from Ucluelet and Tofino, or from the shore in Pacific Rim National Park, near Tofino, during the northerly migration in March and April. Some young gray whales stop here for the summer to feed (for most grays, the full northern migration terminates off western Alaska). They spend their days diving in muddy shallows, eating small shellfish and other organisms, and churning the water to brown.

SPECIAL FEATURES

Harbor seals and California and Steller sea lions can be seen on rocky islets along the coast. The most accessible locales are Race Rocks at the southern tip of Vancouver Island and Sea Lion Rocks off Long Beach in the Pacific Rim National Park. In many other spots along the coast of Vancouver Island, bald eagles prey on salmon and other fish. The eagles take advantage of the best sea views from tall, mature trees. When the tide changes in Johnstone Strait, dozens of eagles divebomb for fish, with the occasional mid-air fight between sparring eagles. Talons locked, they fight over the food until they hit the water.

TRAVELER'S NOTES

When to visit Mar–Apr gray whales (west Vancouver Island); May–Sept orcas, Pacific white-sided dolphins, Dall's porpoises (nth Vancouver Island), harbor porpoises (sth Vancouver Island), gray whales (west Vancouver Island)
Weather Mar–June, often rainy and cold on water; July–Aug, cool on water but usually dry, fog especially mornings on west coast Vancouver Island
Types of tours Half- and full-day tours, multi-day expeditions; inflatables, sailboats, and large whale-watch boats;

some whale watching from ferries
Tours available British Columbia: Alert Bay, Telegraph Cove, Port McNeill, Sointula, Tofino, Ucluelet, Victoria, Nanaimo; Puget Sound area: Anacortes, Bellingham, Friday Harbor, Seattle
Information Tourism British Columbia, Parliament Buildings, Victoria, BC V8W 2Z2, Ph. 1 800 663 6000; Department of Trade & Economic Development, 101 General Administration Bldg., AX-13, Olympia, WA 98504, Ph. 1 800 544 1800

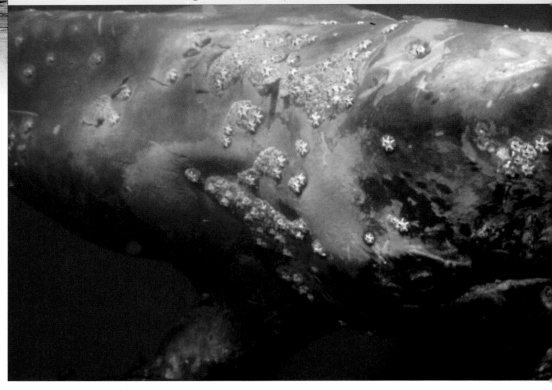

California, Oregon, and Washington

West Coast of North America

A school of blunt-nosed Risso's dolphins (above) seen off the coast of California, west of Monterey.

As regular as the seasons, gray whales come and go on their annual migrations up and down the west coast of North America. With a glassed-in observatory and outdoor overlook, Cabrillo National Monument draws tens of thousands of whale watchers each year who often cheer when gray whales spout. There are numerous lookouts all along the west coast of the continent. Many quiet headlands with good sea views are known only to local people, and whale watching here can be a more contemplative experience. A leisurely driving tour along Highway 1 would allow you to follow the migration of gray whales as they move south or north over a seven-month period.

From late October to December, they arrive off the coast of Washington State from Alaska. Fresh from a summer of feeding in the far north, they are headed for mating and calving grounds in the Baja California lagoons. Off the Oregon coast, the southern migration peaks in the last week of December and the first week of January. Continuing south along the Californian coast, the greatest numbers are seen in the latter half of January. By February, the first gray whales are already starting to appear off southern California on their way back north. The first wave arrives off Oregon in late February and continues through early March. The last stragglers, many of them females with calves, pass along the coast of northern California to Washington in May.

Watching gray whales from land has been a popular activity for decades. For a closer, if colder and wetter look at the action, boat trips depart from some 35 ports along this stretch of coast. Slopping around in the sea alongside these mighty mammals, you can only marvel at their endurance—swimming 12,400 miles (20,000 km) a year back and forth from Alaska to Mexico.

Whale-watching trips encounter various species depending on how far offshore they go and how knowledgeable the operator is. From

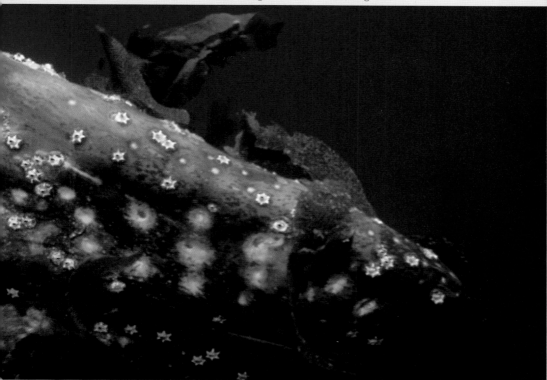

A gray whale (above) swims through kelp off the California coast.
Keen whale watchers (left) in southern California waters.

northern Washington, for example, boat tours trying to catch the last of the gray whales may meet orcas, harbor porpoises, and Pacific white-sided dolphins. Off California, pilot whales, orcas, bottlenose dolphins, and even northern right whale dolphins are sometimes seen.

SPECIAL FEATURES

In the late 1970s, dedicated seabird cruises, taken during August to October to the offshore canyons and banks west of San Francisco and Monterey, began to encounter dolphins and a variety of whales, such as blues and humpbacks. In all, there have been sightings of 26 cetacean species, almost a third of all species, including sporadic sightings of a number of rare beaked whales. Besides blue, humpback, and summering gray whales, the most commonly seen

are minke and fin whales; Pacific white-sided, Risso's, and northern right whale dolphins; and Dall's and harbor porpoises. The whales are here to feed, while many of the smaller cetaceans remain in the area year-round. Other marine mammals are often sighted. This extraordinarily rich marine area extends from the Monterey Submarine Canyon opposite Moss Landing in Monterey Bay to Cordell Bank, some 60 miles (100 km) northwest of San Francisco. It has been protected in three continuous marine sanctuaries—Cordell Bank, Gulf of the Farallones, and Monterey Bay National Marine Sanctuary, the largest, extending along 300 miles (485 km) of coast, to some 50 miles (80 km) out to sea.

TRAVELER'S NOTES

When to visit Dec–May migrating gray whales; June–Sept summering gray whales, north California to Washington; Aug–Oct blue, humpback, and other whales and dolphins off the coast of central California

Weather June–Sept cool to cold on water, even when hot on land; Oct–May, California to Washington, cold, rainy.

Types of tours Half- and full-day tours, multi-day expeditions; inflatables, fishing, sail, and large whale-watch boats

Tours available California: Avila Beach, Balboa, Dana Point, El Granada, Fort Bragg, Hollister, La Mesa, Long Beach, Monterey, Morro Bay, Oceanside,

Oxnard, Point Arena, Redondo Beach, San Diego, San Pedro, Santa Barbara, Santa Cruz, Ventura; Oregon: Charleston, Depoe Bay, Garibaldi, Newport; Washington (west coast only): La Push, Neah Bay, Westport

Information Department of Trade & Economic Development, 1121 L Street, Suite 103, Sacramento, CA 95814, Ph. 1 916 324 5853; Economic Development Department, 595 Cottage Street, NE, Salem, OR 97310, Ph. 1 503 986 0123; Department of Trade & Economic Development, 101 General Administration Bldg., AX-13, Olympia, WA 98504, Ph. 1 360 753 5600

Baja California
West Coast of North America

The blue whale (above), largest living organism on Earth, feeds in the Sea of Cortez in winter.

Baja California, or "lower California," which extends south from the California border with Mexico, is one of the most important areas in the world for whales. In terms of whale watching, there are two dramatically different regions. The first is the open Pacific coast with its sheltered lagoons. Here, friendly gray whales may nudge your boat or ease it up and out of the water, a breathless and rather nerve-racking experience. Don't forget your wide-angle lenses. The other is the Gulf of California, between Baja and the mainland coast of Mexico. This is one of the most species-diverse cetacean and marine areas in the world. Motoring out from the desert coasts of Baja, you feel as if anything could happen.

Baja California first became known as the winter home of the gray whale—a magical place beside the desert where the whales gather to socialize, mate, and raise their calves. The gray whale, which spends part of its year in the lagoons, was once reduced to near extinction. In the mid-nineteenth century, whalers led by Captain Charles M. Scammon discovered the entrance to the Pacific coastal lagoon that would bear his name. The whalers killed almost every gray whale. When Scammon's Lagoon was designated the world's first whale sanctuary in January 1972, gray whales were considered highly endangered. Besides Scammon's Lagoon, the whales are found along the coast in Magdalena Bay and San Ignacio Lagoon. Today, after many years of protection, the gray whales are thought to have recovered fully.

Whale-watching trips to the lagoons can be arranged as part of a 7- to 10-day trip out of San Diego on a self-contained boat. Other visitors drive to Guerrero Negro, near Scammon's Lagoon, and Adolfo López Mateos, at Magdalena Bay, and hire small Mexican boats called "pangas." Land-based whale watchers can stay at several permanent campsites at Magdalena Bay

A gray whale calf breaching (above) in a Baja lagoon. Magdalena Bay (right), an area where gray whales congregate in winter to socialize and breed.

and San Ignacio Lagoon. The idea of sleeping out in the land of the "desert whale" provides a wonderful sense of community, despite the sandy sleeping bags and basic living conditions.

At the south end of the Baja California peninsula, from San José del Cabo and east coast ports, such as Loreto and Bahia de los Angeles, another world opens up—the Gulf of California, also known as the Sea of Cortez. Here you can see fin, humpback, sperm, short-finned pilot, and minke whales, as well as various dolphins including common, bottlenose, and Pacific white-sided. Moreover, it is one of the best places in the world to find blue and Bryde's whales—two species not often seen by whale watchers. Day trips into the Gulf of California are popular, but most people take one- to two-week

boat excursions out of La Paz with a small group of participants. They live aboard the boat and are guided by naturalists who know the area.

SPECIAL FEATURES
In 1993, Mexico's newest reserve, the Upper Gulf of California and Colorado River Delta Biosphere Reserve, was established to protect the vaquita, the small, endangered porpoise that lives only in this part of the world. You may not spot a vaquita but, by way of consolation, humpbacks are sometimes encountered, and fin whales are often seen. Tours leave from the popular tourist town, Puerto Peñasco, in the state of Sonora, just south of the Arizona border. Even though ecotourism is relatively new in this area, it may well become very popular with the excellent birding prospects, the desert and its wildlife, the nearby volcanoes, and diverse cetacean life.

TRAVELER'S NOTES

When to visit *Jan–Apr, gray whales live in the lagoons of Baja California, blue, Bryde's, humpback, fin, and minke whales move into the Gulf of California; year-round, Pacific white-sided, common and various tropical dolphins, bottlenose dolphins (in gray whale lagoons and in the Gulf of California)*
Weather *Dry, clear, warm winters; wind, especially on the Pacific side, can make it cool at sea*

Types of tours *Multi-day expeditions, some day trips; inflatables, pangas, sailboats, and medium-size cruise ships*
Tours available *La Paz, Ensenada, Tijuana, Rosarito, San Diego*
Information *Gobierno del Estado de Baja California Sur, Coordinación Estatal de Turismo, Km. 5.5 Carret. al Norte, Edif. Fedepaz, Apdo. Post. 419, La Paz, Baja California Sur, Mexico, Ph. 52 112 31702*

Hawaii

West Coast of North America

Every Christmastime, humpback whales return to sing, fight, mate, and raise their calves in the tropical blue waters off Hawaii. Humpbacks make their winter homes near these islands, which are checkered with tropical forest, pineapple and coffee farms, and topped with volcanoes. From hotels on the west side of Maui, it is possible to watch the humpbacks as they arrive; they make occasional approaches opposite Lahaina's Front Street, so be prepared.

Migrating from Alaska, the female humpbacks often arrive first, some heavy with calf, others having just given birth. The males arrive later in search of a partner. In the late 1970s, biologists in Maui first witnessed humpback whales fighting. The fighting seems to follow extended bouts of singing among the males who apparently compete to escort and mate with females. Many of the females start these fights or get involved in them by emitting streams of warning bubbles toward ambitious males. Blood can sometimes be seen on the tail or back of a whale or in the water, but the injuries are usually only superficial.

A humpback nearly clear of the water as it breaches (above) close to shore. A spinner dolphin (left) executes a breathtakingly high, spinning leap.

The shallow, warm sea between the islands of Maui, Molokai, and all around Lanai is the prime area in Hawaii for humpbacks and the core area of the Hawaiian Islands National Marine Sanctuary, which was set up mainly to protect these whales. The best access is through boat tours from Lahaina or Kihei on Maui's west side. Maui is a short plane flight from the international airport at Honolulu on the nearby island of Oahu.

The humpback tours usually include sightings of bottlenose dolphins, which often accompany

208

the humpbacks. As in other areas, the dolphins will ride a boat's bow wave, but here they ride the wake of a whale as well. Sometimes, the humpbacks become irritated with the constant companionship, and researchers have seen humpbacks wave their tails at the dolphins, trying to discourage them and drive them away. Beneath the water's surface beside your boat, you can glimpse dolphins—especially if you wear polarizing sunglasses to cut the glare—or even whales swimming, the bubbles from their blowholes leaving a trail through the water.

Most of the other main Hawaiian islands offer whale or dolphin watching and a chance to see a wide variety of tropical cetaceans. Although whale-watch tours focus on the humpbacks, other cetaceans you will see include bottlenose and spinner dolphins, false killer whales, pilot whales, and even the rare beaked whales. From the big island of Hawaii, spinner dolphins—famous for high leaps during which they spin through the air—can be seen throughout much of the year. Other cetacean tours, and more wide-ranging marine nature tours, leave from Oahu and Kauai.

Close behind the powerful tail of a short-finned pilot whale (above). The rugged, volcanic coastline of Kauai (left).

SPECIAL FEATURES

From the main street of Lahaina, signs of whale-watching culture date from the 1970s. In front of the Pioneer Hotel, once home to whaling captains waiting for South Seas postings, about 20 whale-watch boats are berthed, each with its signboard displayed. You can see glass-bottom boats, Chinese junks that look like pirate ships, motor cruisers, Zodiac inflatables, schooners, and trimarans. This is a rare opportunity to go window-shopping for the best whale-watch tour. Key features to look for are an on-board naturalist or nature guide, and a boat equipped with hydrophones, through which you can hear haunting humpback songs.

TRAVELER'S NOTES

When to visit Late Dec–Apr for humpback whales, largest concentrations in Feb and Mar. Year-round: spinner, bottlenose, and pantropical spotted dolphins, false killer whales, short-finned pilot whales, and many others
Weather Warm to hot; sometimes strong Kona winds bring rain and high seas
Types of tours Half- and full-day tours, some extended multi-day

expeditions; inflatables, sailboats, and large whale-watch boats
Tours available Maui: Lahaina, Kihei, Maalaea; Hawaii (big island): Keauhou, Kailua-Kona, Kohala Coast, Honokoha; Oahu: Honolulu, Kaneohe. Kauai: Hanalei
Information Hawaii Visitors Bureau, 2270 Kalakaua Avenue, Suite 801, Honolulu, HI 96813, Ph. 1 808 924 0266

EAST COAST
of NORTH AMERICA

*The East Coast of North America offers some of the world's most
popular ports for whale watching.*

A pristine beach
in Nova Scotia, the Maritimes.

The first commercial whale watching on the east coast of North America occurred in Canada. A summer trip by members of the Zoological Society of Montreal in 1971 took them down the St. Lawrence River to see belugas and large baleen whales. It was so successful that more trips were organized. In 1975, the Dolphin Fleet of Provincetown, Massachusetts, entered the whale-watch business, based largely on humpback whales, and the idea spread up the coast of Maine to the Canadian Maritime provinces and Newfoundland. By the late 1980s, the success of whale watching on the North American east coast eclipsed the West Coast in numbers, and this area leads the world with more than 1.5 million whale watchers every year.

Most of the activity in northeast North America is centered around the feeding grounds of large baleen whales from May to November. It was partly the search by scientists for the mating and calving grounds of some of these whales that led to whale watching in the Caribbean from January to April. In the early 1980s, trips to see Atlantic spotted dolphins had started in the Bahamas, off Florida—the first commercial tours anywhere to offer a chance to swim with wild dolphins. In 1986, the Silver Bank Humpback Whale Sanctuary, north of the Dominican Republic, was established.

Throughout the region, science and education have gone hand in hand with commercial whale-watching development—more so than anywhere else in the world. Photo-identification studies have produced excellent records and catalogs, and databases of sightings have been set up for each baleen whale species in the North Atlantic. If researchers in the Caribbean encounter a certain humpback mother and her newborn calf, researchers in New England might photo-graph them a few months later when the whales come north to feed, documenting

*Watching one of the 3,000 humpbacks
that winter in the Silver Bank Humpback
Whale Sanctuary.*

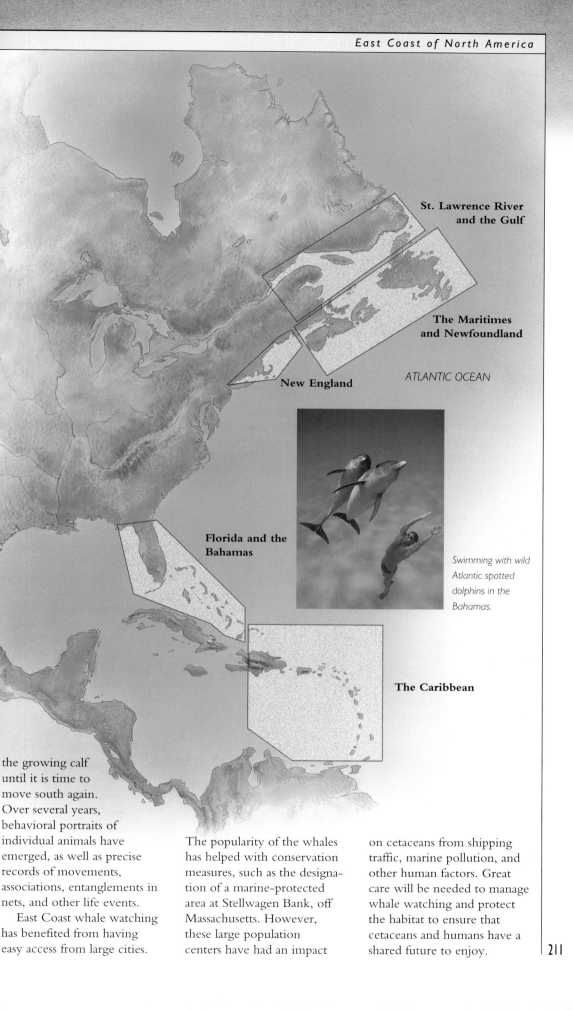

St. Lawrence River
and the Gulf

The Maritimes
and Newfoundland

ATLANTIC OCEAN

New England

Florida and the
Bahamas

*Swimming with wild
Atlantic spotted
dolphins in the
Bahamas.*

The Caribbean

the growing calf
until it is time to
move south again.
Over several years,
behavioral portraits of
individual animals have
emerged, as well as precise
records of movements,
associations, entanglements in
nets, and other life events.

East Coast whale watching
has benefited from having
easy access from large cities.

The popularity of the whales
has helped with conservation
measures, such as the designa-
tion of a marine-protected
area at Stellwagen Bank, off
Massachusetts. However,
these large population
centers have had an impact

on cetaceans from shipping
traffic, marine pollution, and
other human factors. Great
care will be needed to manage
whale watching and protect
the habitat to ensure that
cetaceans and humans have a
shared future to enjoy.

St. Lawrence River and the Gulf

East Coast of North America

I n many places, large whales come close to shore, yet there are few areas where a variety of large whales can be seen far upstream in a river. When the ice thaws in the St. Lawrence River,

A fin whale (left), part of a photo-identification program.

warm-water upwellings from the open sea are carried upstream in submarine trenches. These drive nutrient-rich water to the surface, which results in an explosion of marine life throughout the food pyramid. Whales arrive as soon as the river can be navigated, swimming upstream to take advantage of this wealth of food.

In the province of Quebec, Canada, at the Saguenay-St. Lawrence Marine Park, 230 miles (370 km) upstream from the mouth of the river, belugas and fin and minke whales can be seen. Farther downstream are humpbacks and even blue whales. On the north shore of the river lies Tadoussac, formerly a sleepy summer tourist town and now the bustling main whale-watch center in eastern Canada. Thousands of people pass through here every week in summer, and it is now one of the three largest whale-watch areas in the world, along with New England and the

Canary Islands. Most of the operators in Tadoussac are French Canadian, and some of the shipboard narratives are given in English and French.

From Tadoussac, as well as neighboring towns along the hilly north side of the river, it is possible to take day tours to see mainly fin and minke whales. For the best chance of seeing blue whales, choose a boat from Les Escoumins, 22 miles (35 km) from Tadoussac, or farther downstream. All along the north shore of the river, and in several spots along the south shore, are whale-watch lookouts and additional ports from which other tour companies depart. At the river mouth on the north shore, the Pointe-des-Monts lighthouse provides a superb lookout for blue and other whales.

The dedicated whale watcher, however, will follow the road even farther to the northern Gulf of St. Lawrence, reaching the Mingan Islands near Anticosti Island. The village of Mingan is the home of the Mingan Island Cetacean Study, which has a superb museum and visitor center,

Rugged whale watching (above) in the Gulf of St. Lawrence. A beluga (left) surfaces in the St. Lawrence River.

and a whale-watch and scientific program. Visitors are invited to watch and sometimes assist in blue whale research. This group pioneered photo-identification of blue whales.

The program also focuses on fin, minke, and humpback whales, as well as Atlantic white-sided and white-beaked dolphins, and harbor porpoises. Day tours and week or 10-day excursions are offered. Whale watching is more rugged here, bouncing around in rubber inflatables while dressed in Mustang survival suits (provided by the research team), but it affords a rare chance to see how cetacean research is carried out.

SPECIAL FEATURES

Where the St. Lawrence and Saguenay rivers meet is the southernmost area in the world to see belugas. They can be sighted from the lookout at Pointe Noire or during tours that depart from Tadoussac. The boats don't target the belugas since they are endangered. They are likely to be seen only at some distance. The image of their ghostly white bodies, however, leaves a lasting impression, and if the boat carries a hydrophone, there's a chance of hearing their constant sounds.

Now reduced to about 500 individuals from a population of several thousand, the St. Lawrence belugas have had to endure contaminants and pesticides from the heavy industrial use of the river system, including extensive shipping traffic and effluent outflows upstream from large United States and Canadian cities on the Great Lakes.

Every year, belugas are found stranded and are examined by a dedicated research team. But, there are still some new calves appearing every year and the youngsters, bluish gray beside their white mothers, sometimes poke their heads out of the water or come close to a whale-watch boat to investigate.

TRAVELER'S NOTES

When to visit *St. Lawrence at Saguenay River: June–Nov fin, minke whales, belugas, harbor porpoises; Les Escoumins to Pointe-des-Monts, river and north shore of gulf, especially Mingan Islands: June–Nov fin, minke, humpback whales, occasional orcas, Atlantic white-sided dolphins, harbor porpoises and blue whales in Aug–Nov*
Weather *July–Sept cool to cold on the water, fog common, some heavy summer rains, snow by mid-Oct or Nov*
Types of tours *Half- and full-day tours, some extended multi-day*
expeditions; inflatables, sailboats, and large whale-watch boats; good land-based whale watching
Tours available *St. Lawrence River, Quebec: Tadoussac, Baie-Ste-Catherine, Grandes-Bergeronnes, Les Escoumins, Godbout, Baie-Comeau, Baie-Trinité, Pointe-des-Monts; Gulf of St. Lawrence, Quebec: Longue-Pointe-de-Mingan, Havre-Saint-Pierre; Gaspé Peninsula: Rivière-du-Renard, Gaspé*
Information *Tourisme Quebec, CP 979, Montreal, Quebec H3C 2W3, Ph 1 514 873 2015*

The Maritimes and Newfoundland

East Coast of North America

The Maritimes are the Canadian seaboard provinces of Nova Scotia, New Brunswick, and Prince Edward Island. The Bay of Fundy, between Nova Scotia and New Brunswick, is well known for having the highest tides in the world. This makes for tide rips and tricky currents, especially around the various islands, and contributes to the famous "Fundy fog." These conditions may have discouraged whalers and made the bay a refuge for wildlife.

In August 1980, on a routine marine mammal aerial census of the bay, 19 northern right whales were counted on one day. This led to intensive research and it is now known that there are still about 300 northern right whales in the North Atlantic. Some stay in the Bay of Fundy from August to November and, even though there are so few, sightings are reliable. During this period, the bay becomes a feeding and nursery area for mothers and calves that appreciate the sheltered waters. Access is by tours from southwest Nova Scotia and eastern New Brunswick, especially from North Head on Grand Manan Island. Other adult whales go to open waters in Browns Bank and Roseway Basin south of Nova Scotia. Only

Watchers observe a humpback (left) in Newfoundland waters. A pod of rare northern bottlenose whales (below) swimming in the Gully.

occasional multi-day tours visit this far offshore area.

Humpback, fin, and minke whales are also common in the Bay of Fundy, with sei and sperm whales farther offshore. The western side of the bay is famous for high concentrations of harbor porpoises, which can be seen from land, from ferries to Grand Manan Island, and from whale-watch boats. The porpoises hunt in groups and move rapidly through the water. Unlike dolphins, they tend to avoid boats and do not ride bow waves.

Newfoundland has sheer cliffs from which the whale watcher can view fantastic seabird colonies and abundant whales coming in close to shore. Locals have given unusual names to the common cetaceans. The ever-present harbor porpoises are called "puffing pigs," and dolphins are referred to as "squid hounds." Humpback whales are common

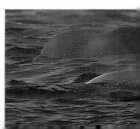

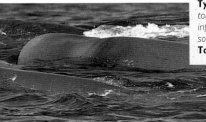

A humpback falls back into the water after breaching (top) in the Newfoundland fog. The small town of Yarmouth (above), in Nova Scotia, is almost encircled by water.

and fin, minke, and long-finned pilot whales are also seen. The Atlantic white-sided and white-beaked dolphins are seen year-round, and less often (mainly in summer) sei, blue, killer, and sperm whales are sighted.

Visit Cape St. Mary's, on the Avalon Peninsula about 125 miles (200 km) from St. John's, for the chance to see humpback, minke, and pilot whales, as well as Atlantic white-sided dolphins. Other good land-based lookouts on the Avalon Peninsula are Cape Race, Holyrood Arm, and Bay de Verde. Besides land-based viewing, there are numerous boat tours all around Newfoundland that venture out to meet the whales.

SPECIAL FEATURES

About 100 miles (160 km) out to sea in the North Atlantic off eastern Nova Scotia, lies the Gully, a deep canyon beloved by northern bottlenose whales (see p. 159). This is the only known place this rare beaked whale can reliably be seen. Proposed as a protected area, the Gully contains large concentrations of plankton, fish and squid. These attract not only bottlenose whales but others, such as blue and sperm whales and, from the warmer waters farther south, several species of dolphin including common, bottlenose, and even striped dolphins.

Of these, northern bottlenose whales are the prize residents and the most commonly seen. There are an estimated 230 bottlenose whales using the Gully—mothers, calves, juveniles, and larger males with their characteristic square heads. Researchers have found their natural behavior a little difficult to study because they are so curious about boats. For the whale watcher who has seen the more common whale species, this provides a new and unforgettable experience.

TRAVELER'S NOTES

When to visit June–Oct fin, humpback, minke whales, various dolphins and harbor porpoises; Aug–early Nov northern right whales, Bay of Fundy and off Nova Scotia
Weather June–Aug cool to cold; fog especially in Bay of Fundy; Sept–Oct colder but often clearer
Types of tours Half- and full-day tours, extended multi-day expeditions; inflatables, sailboats, whale-watch boats; some whale watching from ferries
Tours available New Brunswick:

Grand Manan, Leonardville, Fredericton; Nova Scotia: Halifax, Tiverton, Westport (Brier Island), Cheticamp, Capstick; Newfoundland: St. John's, Bay Bulls, Trinity, Twillingate
Information Tourism New Brunswick, Box 6000, Fredericton, NB E3B 5H1, Ph. 1 800 561 0123; Tourism Newfoundland, Box 8700, St. John's, NF A1B 4J6, Ph. 1 709 729 2803; Tourism Nova Scotia, Box 519, Halifax, NS B3J 2R7, Ph. 1 902 490 5946.

New England

East Coast of North America

Lifted high out of the water, the 16 foot (5 m) black-and-white flippers of the humpback whale, the longest appendages of any animal, can be easily seen above the often white-capped seas of the New England coast. It was here that the humpback whale acquired its name, *Megaptera novaeangliae*, which translates as "great winged New Englander." When whale watching started here in the mid-1970s, humpbacks would occasionally approach boats. Over several years, certain ones got used to the boats and began to interact with whale watchers, leaping repeatedly out of the water and waving flippers and flukes.

The fin whale, seen above surfacing, can grow to 80 feet (25 m); Atlantic white-sided dolphins (below right) at Stellwagen Bank.

Besides humpbacks, more than six species of whale and dolphin—minke, fin, right, and pilot whales, as well as Atlantic white-sided dolphins and harbor porpoises—come to these waters and use the area as a summer feeding ground. The focal point is Stellwagen Bank, a 19 mile (31 km) long submerged sandbank that runs north–south between the tip of Cape Cod and Boston's north shore, at Cape Ann. This massive mound is invisible to whale watchers, but just 65–100 feet (20–30 m) below the sea surface lies a rich ecosystem. Extending across Massachusetts Bay, Stellwagen Bank influences ocean currents and upwellings, both of which help to create high levels of nutrients that support plant plankton and zooplankton, the basis for these important fishing grounds. In 1993, after successful lobbying by whale watchers, an 842 square mile (2,180 sq km) area around the bank was designated a National Marine Sanctuary, an important step in conservation.

The rarest visitors to Stellwagen Bank, indeed the rarest of all whales, are the northern right whales. In April, they sometimes enter Cape Cod Bay and swim so close to shore that they can be photographed against the old clapboard beach houses and traditional homes of Provincetown lining the shore. After April, right whales are seen sporadically in the area and along the Maine coast through October. They are most reliably seen in the Bay of Fundy (see p. 214).

Hook-shaped Cape Cod (above). The distinctive flipper of a humpback whale, Stellwagen Bank (left).

The right whales charge through the water, mouths open, in search of copepods, as do basking sharks in the area, while humpbacks and other whales operate by encircling sand lance, herring and other small fish, and krill. Seeing which whales are present gives biologists an idea of which fish or plankton will be found in high concentrations at the time.

Southern New England is one of the most popular whale-watching areas in the world. With some 30 companies offering tours from 17 communities, whale watching here offers excellent value for money. Most trips carry naturalists and, because of competition, the cost is less than elsewhere. Many trips feature scientists who do their work on board, photographing whales they know by name. More than almost anywhere else, a whale-watch trip to Stellwagen Bank often turns into a grand social occasion in which the naturalist guides and scientists introduce whales by name as they are spotted. They tell stories and exchange gossip about their favorites—who's traveling with whom and which mothers have new calves are the talk of the day.

SPECIAL FEATURES

The Stellwagen Bank National Marine Sanctuary offers much more than whales. Boat trips to the sanctuary and along the coast encounter some 40 species of resident marine birds, including loons, shearwaters, storm petrels, gannets, phalaropes, fulmars, puffins, and murres. Migratory birds also stop here. Harbor and gray seals spend part or all of their year in the region, along with various sea turtles, including leatherback, green, loggerhead, and Kemp's ridley sea turtles.

TRAVELER'S NOTES

When to visit Apr–May humpback, right, minke, fin, and small cetaceans; June–Oct all whales are common, except right whales, which are seen only occasionally; Aug–Oct northern right whales off northeast Maine from Lubec

Weather Warm to hot especially from Massachusetts ports May–Aug, it can be cool out on the sea; rain is most likely in Apr–early May

Types of tours Half- and full-day boat trips; some extended multi-day

trips; large whale-watch boats

Tours available Massachusetts: Provincetown, Nantucket, Barnstable, Plymouth, Boston, Gloucester, Newburyport; New Hampshire: Rye, Hampton Beach, Portsmouth; Maine: Bar Harbor, Kennebunkport, Lubec, Northeast Harbor, Ogunquit, Portland, Boothbay Harbor

Information Massachusetts Office of Travel and Tourism, 100 Cambridge St., 13th Floor, Boston, MA 02202, Ph. 1 617 727 3201

Florida and the Bahamas

East Coast of North America

The Commonwealth of the Bahamas includes more than 700 low-lying islands and 2,000 cays, only 24 of them inhabited. They form a chain extending some 500 miles (800 km) southeast from Florida.

In the mid-1970s, in the waters north of Grand Bahama Island, film-makers discovered that herds of Atlantic spotted dolphins regularly approached boats and photographers in the water. The dolphins appeared so reliably that researchers became interested, and several dive and tour companies began taking people on multi-day tours to meet the dolphins, watch, and sometimes swim with them. Some of the best and most popular tours, organized by an American scientist, helped pay for research into the dolphins' behavior.

The shallow water is relatively calm and extraordinarily clear, the bottom white and sandy, so dolphins can be easily observed in what must be among the best conditions in the world for watching wild dolphins underwater. This has afforded a chance to learn about the dolphins' elaborate behavioral repertoire.

Three dolphin generations on, small groups of Atlantic spotted dolphins still come to boats and swim with snorkelers. The boats typically anchor

Atlantic spotted dolphins (above) are easily seen in the clear, shallow waters.

for several days about 30 miles (48 km) from land on the shallow sandbank, part of Little Bahama Bank, in only 20 feet (6 m) of water. There are an estimated 80–100 dolphins in the population around this sandbank, but only up to 10 are usually seen together. Sometimes bottlenose dolphins travel with the Atlantic spotted dolphins, or they may be seen separately.

The trips are accessible on 3- to 11-day tours from Florida or through West End and Port Lucaya on Grand Bahama Island. They depart aboard dive boats, sailboats, and motor cruisers from May to September, although there are some trips in winter and at other times of the

Spotted dolphins are invited to play (above) with a snorkeler using an underwater scooter. A bottlenose dolphin (right).

year. The dolphins live year-round in these waters, but there is less wind and the seas are generally calmer from May to September. Many of these multi-day trips, often on the way to or from Florida, spend a day cruising in the deeper open waters of the Gulf Stream. There, other cetaceans are sometimes seen, including rough-toothed and other tropical dolphins and short-finned pilot and sperm whales.

In Florida, bottlenose dolphins can be seen from Atlantic and Gulf Coast beaches and sometimes in the intracoastal waterway, along with manatees, diving pelicans, and other wildlife typical of Florida. Particularly good sites for shore-based dolphin watching are in the Florida Keys, the Gulf panhandle, and around Sarasota on the Gulf Coast. For offshore viewing, one long-standing tour operator offers day trips from Key West, about 200–250 days a year, depending on the weather. Inquire locally for information about other operators.

SPECIAL FEATURES

For the past few years, a number of special one-week marine mammal survey expeditions have been offered through Earthwatch. These trips started out mainly as whale surveys but now focus on the flora, fauna, and culture of the northern Bahamas. Some 16 species of whale and dolphin have been recorded. Besides the Atlantic spotted and bottlenose dolphins, the surveys have recorded sperm, humpback, false killer, and short-finned pilot whales, as well as several species of rare tropical beaked whales. The surveys cover the marine habitat around hundreds of islands, from mangroves and shallow sandbanks to deep trenches, so they may yet find more surprises. The trips depart from Great Abaco Island. For information, contact Earthwatch (see below).

TRAVELER'S NOTES

When to visit For Atlantic spotted and bottlenose dolphins year-round; May–Sept best for Bahamas, although some tours at other times of year. Check hurricane reports June–Oct

Weather Warm to hot, generally calm during May–Sept

Types of tours Mainly 3–11 day expeditions to Bahamas (book in advance); some day tours in Florida waters; inflatables, sailboats, motor cruisers, and dive boats

Tours available Florida: Key West; Florida to Bahamas: Jupiter, Dania, Fort Lauderdale, Indialantic, Miami Beach; Bahamas: West End, Port Lucaya, Freeport (Grand Bahama Island)

Information The Bahamas Tourist Office, 255 Alhambra Circle, Suite 425, Coral Gables, FL 33134, Ph. 1 305 442 4867; Florida Division of Tourism, 126 Van Buren Street, Tallahassee, FL 32399, Ph. 1 904 487 1462; Earthwatch, 680 Mt. Auburn Street, Box 403, Watertown, MA 02272, Ph. 1 617 926 8200

The Caribbean

East Coast of North America

The sperm whales lie motionless on the surface, their wrinkled backs exposed. They spout one by one, each explosion followed by a hollow, dull whine as they suck in fresh air. When the whales dive, watchers can see the rest of their bodies in the clear Caribbean waters. The seas of the Caribbean make for idyllic whale and dolphin watching. Most visitors come for the beaches, or for sailing or cruising holidays, but it is also a good place to see a wide variety of cetaceans.

In 1988, the island paradise of Dominica became the first place in the eastern Caribbean to offer whale watching. Trips depart near Roseau and rely on a resident group of 8 to 12 sperm whales found in the deep waters just off the west coast. Besides these, there are pilot whales, false killer whales, and spinner and spotted dolphins. Pygmy sperm whales, rarely seen anywhere else in the world, are often found in the warm waters around Dominica. Bottlenose, Risso's, and Fraser's dolphins are also encountered, as well as

Bottlenose dolphins love to ride a boat's stern wave (left). The humpback whale (below right) is tail-breaching in Silver Bank Humpback Whale Sanctuary.

orcas, dwarf sperm whales, and melon-headed whales. Dive-boat operators on Martinique and Guadeloupe, the main islands near Dominica, now offer tours to see sperm and other whales.

In the eastern Caribbean, there is excellent spinner and spotted dolphin watching off St. Vincent and the Grenadines. Based out of Arnos Vale, on the big island of St. Vincent, the tours are almost year-round, although mid-December to mid-February can become windy. Farther south, Grenada has several dolphin-watch operators who take visitors to see spinner, spotted, and bottlenose dolphins, with a chance to see short-finned pilot and sperm whales. The Kido Project in Carriacou offers youth environmental projects that include dolphin watching on a catamaran—a chance for young people to become inspired and motivated by encounters with cetaceans.

The squared head shape of the three sperm whales (above) surfacing together is unmistakable. Clear sea and lush vegetation (right) make the Caribbean a popular destination.

Some 3,000 humpbacks gather in the relaxed, warm-water setting of the Caribbean, and there's a good chance of seeing tail-lashing females and singing males. The key area is the Silver Bank Humpback Whale Sanctuary, north of Dominican Republic. The whales also gather on Navidad Bank in Samaná Bay, on the west side of Puerto Rico, and north of the British and American Virgin Islands. Whale watchers who have met individual humpbacks on the feeding grounds off New England and eastern Canada may encounter the same animals in the Caribbean. North Atlantic humpbacks use the Caribbean for singing, mating, and calving, much as the North Pacific humpbacks use Hawaii, Mexico, and southern Japan. In winter, the tropical seas around these areas resound with singing humpback whales.

SPECIAL FEATURES

Since the early 1990s, Paul Knapp Jr. has invited several hundred people each winter (Dec 30– Apr 15) to listen to male humpback whales singing in the waters north of Tortola. Visitors concentrate on sounds delivered by high-quality hydrophones and speakers. Even when the whales are some distance away, well out of sighting range, the sounds coming into the boat can be quite loud. No need to travel long distances, or wait for calm seas. Just slip into a comfortable boat, drop a hydrophone over the side, lie back and listen. Such a "whale-listening" tour gives a new dimension to the experience of whale watching.

TRAVELER'S NOTES

When to visit *Jan–Apr humpback whales; year-round sperm whales and various dolphins; inquire locally as tour operating seasons vary (for example, St. Vincent, dolphins, Apr–Sept)*

Weather *Warm to hot, seas sometimes rough; tours often confine activities to the lee side of islands. Storms sometimes Aug–Oct; rainy season varies, but seasonal daily rain may last for only part of a day*

Types of tours *Half- and full-day tours, some extended multi-day expeditions in Dominican Republic; inflatables, sailboats, and large whale-watch boats*

Tours available *Dominican Republic: Samaná, Puerto Plata, Santo Domingo; Puerto Rico: Rincon; US Virgin Islands: Long Bay, St. Thomas; British Virgin Islands: Road Town, Tortola; Guadeloupe: Le Moule; Dominica: Roseau; Martinique: Carbet; St. Vincent: Arnos Vale; Grenada: St. George's, Carriacou (Grenadines)*

Information *Caribbean Tourism Organization, 80 Broad Street, 32nd Floor, New York, NY 10004, Ph. 1 212 682 0435*

SOUTH AMERICA

*South America has everything from river dolphins deep in the jungle
to the great southern right whales off Patagonia.*

Galápagos
Islands

Commercial whale-watching tours in South America started in Argentina in 1983. As early as 1970, Roger Payne began his photo-identification work on southern right whales at Península Valdés, Patagonia. Many whale and dolphin researchers from around the world made treks to this area through the 1970s and early 1980s, and their fascinating stories of Patagonia and the whales and dolphins living in the sheltered bays of the peninsula spread worldwide.

Even before whale study and research began in Patagonia, cetacean watching was a part of jungle excursions to the Amazon as well as boat tours around the Galápagos Islands. But the success of whale watching in Argentina has led to the expansion of dolphin watching along the coast of Brazil and in the Amazon basin, and the initiation of new whale-watch tours along the coasts of both Ecuador and Colombia.

All the other countries in South America, except land-locked Paraguay, have at least

A southern right whale starts to breach off Patagonia, Argentina.

In the waters off the Galápagos Islands live dolphins and other marine life.

some cetacean-watching activity, and even where there are no commercial tours, there are still viewing opportunities from land. Wherever you go, along the coast and up rivers, keep your eyes open.

Many of the large whales are visitors to South America. They spend part of their year feeding in the Antarctic waters and then swim north to warmer climates, heading either east or west of Tierra del Fuego to reach the South Pacific or South Atlantic. In most cases, the migration

stops south of the equator, but certain humpbacks cross the equator and move as far north as Colombia and Costa Rica—the longest migration by an individual whale as yet documented.

South American waters provide mating, calving, and nursery grounds for the large whales, while the smaller whales and dolphins live here year-round and find both their food and mates in the same waters. Southern South

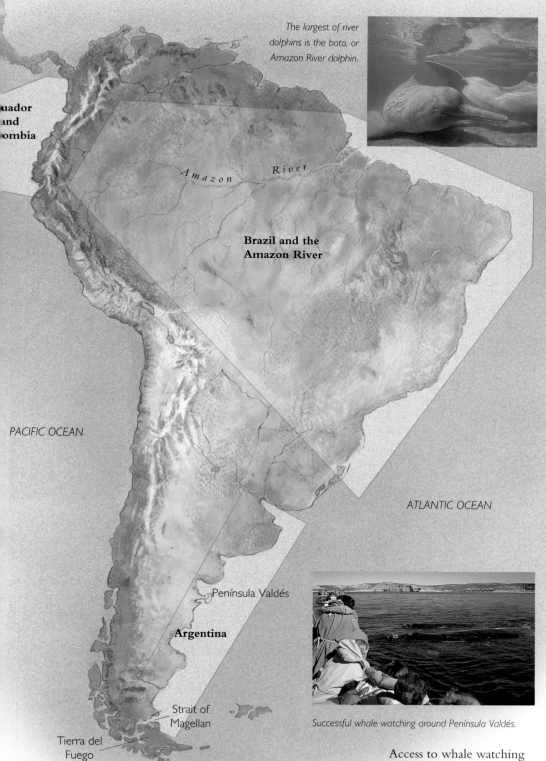

The largest of river dolphins is the boto, or Amazon River dolphin.

Ecuador
and
Colombia

Amazon River

**Brazil and the
Amazon River**

PACIFIC OCEAN

ATLANTIC OCEAN

Península Valdés

Argentina

Strait of
Magellan

Tierra del
Fuego

Successful whale watching around Península Valdés.

America has a great diversity of unusual species: at least six dolphins and porpoises are found mainly or only in these waters, especially off Chile and Argentina. As well, some of the rarer beaked whales are seen in the offshore waters or stranded on the beaches of southern South America, including two new species of beaked whale described as recently as the early 1990s.

Access to whale watching in this area is more difficult than in North America and some other parts of the world. But the thrill of seeing river dolphins, the big southern rights, or rare dolphins in and around the Strait of Magellan make this a unique and interesting region to visit.

223

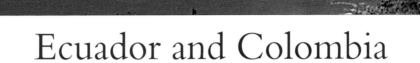

Ecuador and Colombia

South America

Whale watching in the northwest corner of South America is hot and steamy, and offers the chance to meet some rare tropical species as well as migrating whales. Ecuador and Colombia are relatively new to whale-watch tours. Ecuador has long been famous for its offshore Galápagos Islands, and Colombia has a long coast on the Caribbean, but the best whale watching in Ecuador and Colombia occurs along the Pacific coast. The main attraction is Southern Hemisphere humpback whales, which cross the equator on their extraordinarily long migration. They migrate from the Antarctic, where they feed in the austral summer, to Colombia and Ecuador, where they breed—a round trip of 10,351 miles (16,668 km).

In Colombia, humpback research is centered around the island of Gorgona, which is located 35 miles (56 km) offshore from the mainland. During the whale-watch season from August to October, the best place to depart for on-the-spot tours is from the mainland at Juanchaco. Some of the whale-watching boat owners lease cabins on

Humpback whales charging through the water off Gorgona Island.

the beach at Juanchaco and also offer package tours that depart from Cali and Buenaventura. Besides the humpback whales, it is possible to see various dolphins, including bottlenose, as well as orcas and false killer whales.

In Ecuador, whale watching of seasonal humpback whales is offered from June to mid-September along the mainland coast out of Machalilla National Park and the fishing ports of Puerto López and Salango. Many whale-watch trips head from the park to the island of La Plata, which is an excellent area for humpback whales. For land-based whale watching, the Hotel Punta

Carnero, at Punta Carnero, can offer superb views, from a viewing platform or some of its seaside hotel rooms, of humpback whales spouting, lifting their white flippers, and breaching.

The best time in Ecuador for other cetaceans is December to May, with sun and calm seas on most days. Bottlenose and pantropical spotted dolphins patrol the inshore waters. Farther out to sea, look for sporadic visits by Bryde's and false killer whales; orcas; striped, common, and Risso's dolphins.

There is reliable bottlenose dolphin watching in the estuary of the River Guayas in the Gulf of Guayaquil. Some 400 to 500 dolphins live in the river year-round, but the best months to feel soft breezes and have a dozen dolphins splashing beside your boat are June through November.

Gorgona Island (above), where researchers study humpbacks. The primeval Galápagos landscape (left).

SPECIAL FEATURES

The Galápagos Islands, located 600 miles (975 km) offshore in the equatorial Pacific, were one of the first international ecotourism destinations. The Galápagos Marine Reserve dates from 1959 and the islands were opened to organized tourism in 1970. Today, all the waters around the Galápagos are a whale and dolphin sanctuary.

Visitors can sometimes see bottlenose, common, and spinner dolphins near the islands. Farther offshore are sperm whales and Bryde's whales, but it is usually necessary to take overnight tours to see them. Various international operators take people on nature tours that include dolphins and the odd whale as part of the natural history experience, but few trips feature cetaceans.

Even cetacean-oriented tours include visits to see the extraordinary range of island habitats from lava deserts to tropical forests, to encounter the amazing endemic species that include marine iguanas and giant tortoises, and to visit the Charles Darwin Biological Research Station.

TRAVELER'S NOTES

When to visit Year-round: bottlenose, pantropical spotted, and other dolphins Ecuador and Galápagos; offshore Ecuador: spinner dolphins, orcas, sperm, and Bryde's whales; June–Sept humpback whales Ecuador, and Aug–Oct Colombia

Weather Coastal Ecuador and Colombia hot, humid year-round, sometimes rainy during humpback whale season; the Galápagos are drier, best months for sea conditions Mar–Aug, although year-round possible

Types of tours Half- and full-day tours, extended multi-day expeditions;

inflatables, sailboats, small motorboats, and small cruise ships

Tours available Ecuador: Guayaquil, Quito, Machalilla National Park, Puerto López, Salango; Colombia: Cali, Buenaventura, Bahía, Juanchaco, Ladrilleros, Bahía Solano, El Valle, Chocó

Information Fundación Ecuatoriana para el Estudio de Mamíferos Marinos (FEMM), Velez 911 y 6 de Marzo, Ed. Forum, 5to. piso, Of. 5-16, PO Box 09 01 11905, Guayaquil, Ecuador, Ph./Fax: 593 4 524 608; Estación Científica Charles Darwin, Isla Santa Cruz, Galápagos, Ecuador

225

Brazil and the Amazon River

South America

Watching pink dolphins with long beaks splash and chase fish through the brown, muddy water of the Amazon is a strange sight, even on a jungle safari. Many local and native Indian people thought botos, or Amazon River dolphins, brought bad luck and tried to avoid them, and this reputation has effectively protected the dolphins. Although botos are not dangerous, they are highly curious and playful. People in boats report that the dolphins grab the paddles in their mouths and even rub their backs on the undersides of boats.

The tucuxi (above) is found along the Brazilian coast and in the Amazon River region.

Botos are found only in the tropical rain forest of South America. This covers much of the vast Amazon and Orinoco river basins that drain more than half of South America, including parts of Brazil, Bolivia, Peru, Ecuador, Colombia, Venezuela, and Guyana. You are able to see botos in all of these countries. The best ports to depart from include Leticia and Puerto Nariño in southern Colombia; the Pacaya-Samiria Reserve in eastern Peru; and Manaus and various other ports on the Brazilian Amazon with its many large tributaries.

Many cetacean watchers have heard of the boto, but few realize that there is also another dolphin that is seen on river-dolphin and rain-forest excursions. It's called tucuxi. Classed as an oceanic dolphin, its range includes the Amazon and Orinoco rivers. The difference between the two dolphins is evident when you see them together. The tucuxi looks sleek and streamlined, while the pinkish boto is an awkward-looking, angular animal, with huge, broad flippers, a beak full of tiny teeth, and pinholes for eyes.

The tucuxi is most commonly found living along the Atlantic coast of Brazil, and is the most popular cetacean for whale and dolphin watchers in Brazil. It can be seen on sailboat tours from the north-central coast of Santa Catarina Island. Other cetacean tours in Brazil focus on spotting southern right whales from the cliffs in a breeding and calving area near the south end of Santa Catarina Island; humpback whales in the

Boto (above) spotted in the waters of the Amazon River. A river safari through the jungle (left) uses the Amazon and its tributaries as a highway.

fishery, which uses traditional hand nets, supports more than 100 families. For long periods, some 25–30 dolphins in various groups help the fishers by driving the fish closer to shore and into the nets. The dolphins benefit by being able to take some fish as well.

In 1993, Laguna declared the lagoon an ecological sanctuary for dolphins. In the future, Laguna may offer tours for the southern right whales, which come close to shore near Imbituba between June and September, but the right whales as well as the dolphins are best and most easily observed from land.

National Marine Park of Abrolhos, some 25 miles (40 km) offshore from southern Bahia; and spinner dolphins at Fernando de Noronha archipelago in Pernambuco. Bottlenose dolphins can also be seen in many places, including year-round from the southern coast of Santa Catarina.

SPECIAL FEATURES

In Santa Catarina, some distance south of the main southern right whale area on Santa Catarina Island, you may witness rare and extraordinary cooperation between humans and dolphins. Some 200 bottlenose dolphins reside inshore around the mouth of a large lagoon near the coastal city of Laguna. The dolphins come close to shore year-round, except during July and August, for mullet fishing. The mullet

TRAVELER'S NOTES

When to visit *Year-round: river dolphins in Amazon-Orinoco but best during low-water seasons as dolphins are more confined, and you avoid rainy season; tucuxi dolphins at Santa Catarina Island, spinner dolphins at Fernando de Noronha archipelago, both in Brazil. June–Sept/Oct: southern right whales southern Santa Catarina Island. June–Dec humpbacks in National Marine Park of Abrolhos in Brazil*

Weather *Amazon-Orinoco hot, humid conditions; in southern Brazil, whale season is cold, windy, even from lookouts, often rough on the water*

Types of tours *Half- and full-day tours, some extended multi-day expeditions; inflatables, sailboats, motorboats, canoes, and river ferries; some watching from land*

Tours available *Brazil: Florianópolis, Caravelas, Manaus; Colombian Amazon: Bogotá, Leticia, and Puerto Nariño; Peru: Iquitos*

Information *Oceanic Society Expeditions, Fort Mason Center, Building E, San Francisco, CA 94123, Ph. 1 415 441 1106; International Wildlife Coalition/Brasil, C.P. 5087, 88040-970, Florianópolis, SC, Brazil, Ph./Fax 55 48 234 1580*

Argentina

South America

Fabled Patagonia, celebrated by explorers,
biologists, and modern writers, is the
remote southern region of Argentina.
Called by one writer "the uttermost part of the
Earth," Patagonia comprises the provinces of
Chubut, Río Negro, Santa Cruz, and the
territory of Tierra del Fuego.

Patagonia is a semiarid tableland full of sheep
farms, mineral deposits, and vast open areas.
Much of the coastline is stark, windswept cliffs,
but halfway down the long Patagonian coast, a
large peninsula, called Península Valdés, juts out.
This provides two large gulfs and, within them,
some sheltered bays that are favored by the rare
southern right whale.

In the 1970s, Patagonia became famous in the
cetacean world when scientist Roger Payne (see
p. 96) moved his family to a cliff-side house to
study right whales. He and later researchers
developed the individual photo–identification
method for these whales, and he recorded their
sounds and most intimate behavior.

Arriving every July, southern right whales use
the waters around Península Valdés to mate,
calve, raise their young, and sometimes play. As
Payne and his family witnessed, the afternoon
show begins as the winds reach 20 miles (32 km)
per hour off the Patagonian coast. One by one,
the southern right whales raise their broad tails

Tail high, a southern right whale (above) "sails" across the bay in
the sheltered waters around Península Valdés, Patagonia.

high out of the water and "sail" across the bay;
then they dive and return to repeat the
maneuver. To see one of these huge tails sailing
past from a small boat, is an awesome sight.

Day tours to watch southern right whales
leave from the towns of Puerto Madryn, Trelew,
and Puerto Pirámide; for seven-day or longer
package tours, book ahead through international
tour operators. All tours go through, or leave
from, Buenos Aires. Package tours can include
visits to elephant seal, sea lion and seal rookeries,

An orca (above) teaches her calves how to catch sea lions, then return to the water. A whale-watch boat (left) at Puerto Pirámide.

Magellanic penguin colonies (at nearby Punta Tombo), as well as land-based excursions into Patagonia itself.

Península Valdés is the place to witness acrobatic dusky dolphins and the drama of orcas attacking sea lions. Orcas feast on them during the pupping season from mid–February to mid–April. At Punta Norte, on the other side of Península Valdés, there is a special viewing area where you can see the orcas patrolling the shallows and charging right out of the water onto the beach to pick off young sea lions.

can be seen on dolphin–watch tours out of Puerto Deseado in the far south of Argentina. These dolphins appear in the same season as the southern right whales, but spring weather is poor in these far southern waters and the best viewing season is December to March. This overlaps with part of the period when orcas grab the young sea lions off the beach at Península Valdés. Whale watchers who miss the right whale season at Patagonia can see Commerson's dolphins and orcas instead. It is also possible to see indigenous Peale's dolphins, but to spot some of the other rare small cetaceans found only off southern South America, you need to join a long-range cruise which includes the Strait of Magellan.

SPECIAL FEATURES
Many experienced cetacean watchers long to spend time in southern Patagonia. Several cetacean species are found only around the tip of southern South America (including the waters of either or both Argentina and Chile). The best studied of these is Commerson's dolphin, a small black-and-white dolphin, which

TRAVELER'S NOTES

When to visit Mid-July–Nov southern right whales at Península Valdés, Patagonia, best Sept–Oct; orcas year-round but catch sea lions on beaches mid-Feb to mid-Apr; Dec–Mar Commerson's dolphins and Peale's dolphins near Puerto Deseado
Weather Mid-July–Nov cold during right whale season, cool at sea even on best days for dolphin watching in Dec–Mar
Types of tours Mainly extended multi-day expeditions, some day tours

locally; inflatables, fishing boats, sport-fishing boats, sailboats, kayaks
Tours available Buenos Aires; Chubut province: Puerto Pirámide, Puerto Madryn, Rawson, Trelew; Santa Cruz province: Puerto San Julian, Puerto Deseado
Information Oficina Informática Turística (Tourism Information Office), Avenida Santa Fe 883, 1059 Buenos Aires, Argentina. Ph. 54 1 312 2232 or 312 5550

EUROPE *and* AFRICA

Greenland

The spiritual home of dolphin watching since the days of Aristotle, Europe and Africa are excellent for land-based, scientific, and educational whale watching.

Cornish coast (above); a dolphin off the coast of Cornwall (above right).

Commercial cetacean watching in Europe and Africa started in 1980, when a Gibraltar fisherman began offering boat tours to see three species of Mediterranean dolphin. By the mid-1980s, trips were being offered in France, Britain, Ireland, and Portugal to see dolphins. At the same time, interest in land-based watching of southern right whales was growing on the South African coast, but it wasn't until the end of the 1980s that whale and dolphin watching really took off there. In addition, substantial industries started up in Norway, Italy, and the Canary Islands followed closely by the Azores.

Whale watching for sperm whales and various baleen whales out of Andenes, in northern Norway, is in many ways a model for successful ecotourism. Experienced naturalists, or nature guides,

and scientists work on board whale-watch boats, and a museum and whale center are located in the town. Watching whales here has become a learning experience as well as simply good fun.

In Italy, guided fin whale tours out of Porto Sole and San Remo, near Genoa, have become popular. In the Azores, there is an amazing diversity of cetaceans and both land- and boat-based whale watching.

The Canary Islands, off southern Morocco, have had the fastest growth in whale watching of anywhere in the world. Between 1991 and 1996, the number of people watching whales increased from fewer than 1,000 to an estimated 500,000 visitors a year taking trips on at least 46 registered boats, owned by some 31 companies. This

makes the Canary Islands one of the three largest whale-watch destinations in the world, along with southern New England and the St. Lawrence River. There are outstanding opportunities for observing local short-finned pilot whales and various dolphins. However, few operators carry naturalists on board, so make enquiries about who the responsible operators are.

Aside from the Canary Islands, Europe and Africa have been responsible for some of the finest whale-watch tours. In South Africa and Ireland, in particular, thoughtful and well-developed walking tours have given participants a chance to see whales during guided tours along the cliffs.

From Italy, Greece, Croatia, and France, along with some other countries, participants may sign up for 3–10 days to accompany scientists on high-quality tours

A dolphin (left) swims and leaps alongside watchers aboard an inflatable tour boat on the Mediterranean.

Iceland

The
Faeroes

Norway

Britain
and
Ireland

Canaries and
Azores

The
Mediterranean

South Africa
and
Madagascar

*Searching (above) around the Azores
for sperm whales, just one of many
cetacean species sighted in this area.*

that contribute to the conservation of cetaceans.

Yet the newest and perhaps
most exciting area for whale
watching in the world is
Iceland, where whale watchers
began visiting in numbers in
1995. Tourists are now
arriving from all over Europe,
North America, and even
farther afield, to watch and
photograph whales against a
spectacular background of
volcanoes, icefields, and the
midnight sun.

Iceland

Europe and Africa

At an hour past midnight, in the full sun of early summer, minke whales are spouting and terns diving—all scooping and swallowing tiny fish. Lulled by flat, calm seas in the shadow of snow-capped peaks off northeastern Iceland, a boatload of whale watchers sights a pod of playful whales, their spouts and broad backs gleaming in the sun.

A minke whale (above) is bathed in the warm glow of the midnight sun as it surfaces near Húsavík, northern Iceland.

Lying just below the Arctic Circle in the far North Atlantic Ocean, Iceland is roughly midway between North America and Europe. Icelandair, Iceland's national airline, offers one- to three-night stops in Iceland for Europeans or North Americans crossing the Atlantic, at no extra charge. To date, no other scheduled airlines stop in Iceland.

More than ever, people today are headed toward Iceland for whale-watching and nature excursions. Whale-watching sites have opened up in various villages and towns around the country. Besides minke whales, fin, killer, sperm, sei, humpback, and blue whales can be seen, along with Atlantic white-sided dolphins, white-beaked dolphins, and apparently shy harbor porpoises.

Although midnight whale watching in late June can be enjoyed from various parts of the country, Húsavík, a fishing village in northeastern Iceland, has become the center for whale watching. Friendly minke whales provide the main interest during the whale-watch season. Humpback whales are sometimes encountered, and, several times every year, blue whales (normally feeding farther offshore) come into the bay and feed for a few days or a week. Whales are sometimes seen from the town and, by boat, often within an hour of leaving the dock.

At Höfn, in southeast Iceland, where whale watching started in 1991, whale watchers board a 150 ton (135 tonne) lobster fishing boat to see humpback and minke whales, as well as harbor porpoises and sometimes orcas. On the west coast, 10–12 miles (16–19 km) offshore from Stykkishólmur, blue and sperm whales can be

Iceland offers visitors a wealth of wildlife, such as the humpback whale (above) breaching, and a close-up view of a minke whale for whale watchers near Húsavík (right).

spotted on full-day trips that include visits to thriving seabird colonies and also to the famous Snaefellsjökull glacier, location for the movie *Journey to the Center of the Earth.*

Other whale-watch ports around the country include Grindavík, Keflavík, Arnarstapi, Ólafsvík, and Dalvík. Some of the operators, such as at Húsavík, use sturdy wooden Icelandic fishing boats customized for whale watching, while others use newer fishing boats.

Access for all tours is through the international airport at Keflavík, with a transfer to the national airport in nearby Reykjavík for ports in northern Iceland, including Húsavík. Whale watching is possible on day tours, while some visitors come as part of a guided package tour. Such trips can be designed around whale-watching activities or might include other nature and cultural tours.

SPECIAL FEATURES

In northeastern Iceland, near Húsavík, there is superb bird watching in the Lake Myvatn area. Species include some rare ducks. There is regular volcanic activity nearby, which can be seen in hot, sputtering mudholes and a lunar-like landscape. It was here that American astronauts trained for their moon walks.

From Höfn, on the southeastern coast, there are excursions by skidoo or snowmobile across Vatnajökull, Europe's largest icecap. In the southwest, from Grindavík, close to Reykjavík, tours visit Eldey Island, the last stand of the great auk—a flightless marine bird that lived on the island before being driven to extinction. Eldey Island is now the site of a huge gannet colony.

TRAVELER'S NOTES

When to visit *May–Sept several cetacean species (humpback whales more in early summer, orcas in late summer); best period June–Aug*

Weather *Cold on the water; rain and rough seas intermittent; snow is possible early and late in whale-watch season*

Types of tours *Half- and full-day tours,* some extended multi-day expeditions; fishing boats and large tour boats

Tours available *Húsavík, Höfn, Dalvík, Hauganes, Stykkishólmur, Keflavík, Grindavík, Arnarstapi, Ólafsvík*

Information *Icelandic Information Center, Bankastraeti 2, 101 Reykjavík, Iceland, Ph. 354 562 3045*

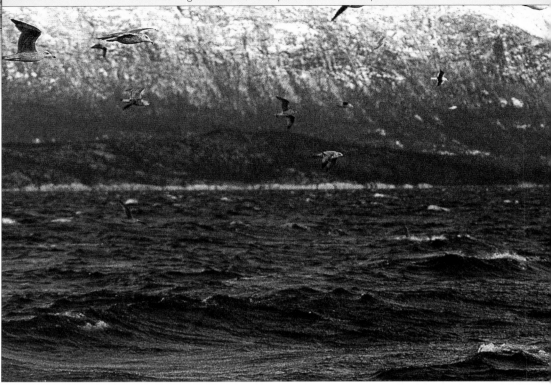

Norway

Europe and Africa

In the rugged seas off northern Norway, sperm whales break the water, their long backs sliding up and down the waves like logs adrift. These sperm whales, all males, come here to feed on squid. Sperm whales are the largest of the toothed whales, and have the biggest brains of any animal. Almost certainly a part of that brain power is employed in their complex sensory systems, which enable them to hunt and catch squid in the blackness of submarine canyons.

Guillemots and kittiwakes nesting on an island (left) in the Norwegian Arctic.

The colorful fishing port of Andenes, which has its own airport (accessible through Oslo), is the most northerly place in the world from which to see sperm whales; it is also one of the best. Here, the continental shelf is closer to shore than anywhere else in Norway, little more than two hours away by boat. The four- to five-hour tours are offered twice a day from the end of May to mid-September.

During the summer months, Andenes—located on Andøy Island, which is connected to the mainland by bridges—is awash with visitors from some 40 countries, most coming for whale-watching trips that may be the most culturally diverse in the world. The brochures are printed in three or four languages, and multilingual naturalists travel on each trip. The naturalists are keen Scandinavian scientists who do some of their photo-identification and other work on the boats. In this way, whale watching in Norway has significantly contributed to the development of cetacean field work in northern Europe.

Besides sperm whales, other cetaceans sometimes encountered include minke, fin, and pilot whales, white-beaked dolphins, orcas, and harbor porpoises. In June and July, daylight lasting almost 24 hours allows time to appreciate every aspect of nature—and to find the whales. Some 95 percent of trips successfully spot whales, and the main tour operator guarantees that if whales are not seen, a free trip will be provided.

South of Andenes along the Norwegian coast, in an area called Tysfjord, the military-like formations of orcas herding herring close to

An orca (above) in Norwegian waters. On day tours (left) from Tysfjord, northern Norway, sightings of orcas are guaranteed.

on Bleiksoy Island, is just off the western side of Andøy. Two-hour tours visit the colony, where 70,000–80,000 pairs of puffins, 6,000 pairs of kittiwakes, and numerous cormorants, storm petrels, common guillemots, and fulmars breed.

Two of several local museums of special interest to nature lovers are Hisnakul—a new museum that covers local bird life, fisheries, the moors, geology, and the northern lights—and the Andenes Whale Center, a museum of whales, whale research, and north Norwegian whaling. At the Whale Center, there are slide shows and daily public lectures on whale behavior, research, and ecology. In addition, visitors can obtain a memento of their whale-watch trip, in the form of a diploma and badge, by joining the Royal International Whale Safari Club. Membership funds are used to support the research and information activities of the Whale Center, which is affiliated with the whale-watch operation in Andenes.

shore every autumn lend a touch of drama to the dark, moody Norwegian fiords. The prime viewing period is short (10 October to 20 November), with daylight extending to only four to six hours a day at that time. Sometimes the orcas can be seen from land but almost always within an hour or two of departure from the dock. The popularity of these tours caused traffic jams on local roads and in the fiords in recent years, but operators and researchers are trying to improve the situation. Access is through Narvik Airport, about an hour and a half away.

SPECIAL FEATURES

In the Andenes area, on or near Andøy Island, visitors are offered a wide range of activities. One of Norway's better bird cliffs,

TRAVELER'S NOTES

When to visit *Late May–Sept in the Andenes area for sperm and other whales and dolphins; Oct–mid-Nov around Tysfjord for orcas*
Weather *Cold on the water late May–Sept in Andenes and Oct–Nov in Tysfjord; Tysfjord has very short days (few hours daylight by mid-Nov)*

Types of tours *Half- and full-day tours, some extended multi-day expeditions; inflatables, sailboats, and large boats*
Tours available *Andenes, Nyksund, Myre, Stø, Storjord*
Information *Andøy Reiseliv, Tourist Information, P.O Box 58, N-8480 Andenes, Norway, Ph. 47 76 11 56 00*

Greenland and the Faeroe Islands

Europe and Africa

Greenland, the second largest island in the world after Australia, is made up mostly of an icecap up to 14,000 feet (4,300 m) thick. Greenland's 55,900 residents are mainly Inuit people who arrived from North America around 2,500 BC, along with Norse settlers, who followed after Eric the Red's landing in AD 985. A self-governing territory of Denmark, Greenland today remains more remote and expensive for local travel and amenities than adjacent northern Canada to the west, or Iceland to the east, but its whale watching has a character all of its own. Tiny, often colorful villages built on icy fiords offer superb baleen and toothed whale watching. The island's deeply indented coast provides shelter for mariners inside icy bays with craggy shores. Greenland has a productive marine environment that attracts plankton and other invertebrates, fish, seals, dolphins, and whales.

The most popular summer site to see whales is out of Paamiut (Frederikshåb) on the relatively warm southwest coast. Humpback whales come here to feed and, with any luck, you can get a photograph of them lifting their giant flukes with a drifting iceberg or glacier as the backdrop. Greenland humpbacks are arguably the most energetic or adventurous humpbacks in the North Atlantic. These whales come from the Caribbean, passing through feeding grounds in

Tourists in small inflatable boats check out a tide-water glacier (above) around Prince Christian Sound, in southern Greenland.

the Gulf of Maine, the St. Lawrence River, and off Newfoundland to feed around Greenland's waters. Their choice of location for summer feeding is probably determined by the areas they were first taken to as calves.

Besides humpbacks, tours out of Paamiut often encounter fin and minke whales feeding in large numbers in areas with massive plankton blooms. After a good feeding bout, whales can frequently be seen leaping and cavorting, and some may even come over to your boat to investigate, rolling on their sides at the surface or spyhopping to catch a glimpse of the whale watchers.

North along the west coast of Greenland, from the Disko Bay area, there are boat charters,

A humpback breaching (above). A flotilla of fishing boats (left) moored in Klaksvik, Faeroe Islands.

as well as kayaks for rent, to see whales close up, especially at Aasiaat and Ilulissat (Jakobshavn). Imagine kayaking with fin whales and puffing porpoises moving alongside you at water level. In September and October, belugas and fin whales may be sighted. Although the days are short, for whale watching, you might find narwhals farther offshore.

The Faeroes comprise 18 islands connected by bridges and small ferries. From these Danish self-governing islands, located between Iceland and Scotland, dolphins and sometimes whales can be seen on boat tours and from shore. Classic Faeroese Viking boats offer bird-watching and

nature tours. Although the gregarious pilot whales are hunted here, the marine nature opportunities are still outstanding. Many visitors ask to watch whales and dolphins, to visit steep seabird cliffs, and to see local seals.

SPECIAL FEATURES

Just 28 miles (45 km) east of Ilulissat (Jakobshavn) is the most active glacier in the Northern Hemisphere, producing an estimated 20 million tons (18 million tonnes) of ice every day. Greenland Tourism offers a range of nature, glacier, cultural, and historical tours by helicopter, boat, kayak, and dogsled. Don't miss the seabird cliffs in spring and early summer on Mykines and other Faeroe islands. Access is through Tórshavn.

Fin whales (above) spotted off the coast of Greenland, where they feast on plankton.

TRAVELER'S NOTES

When to visit June–Aug Greenland; Sept–Oct whales common but weather can be poor; May–Oct Faeroes
Weather Cold on the sea, rain and snow a possibility
Types of tours Half- and full-day tours, some extended multi-day expeditions; kayaks, fishing boats, large touring boats
Tours available Greenland: Paamiut (Frederikshåb), Aasiaat, and Ilulissat

(Jakobshavn) in Disko Bay, Ammassalik; the Faeroe Islands: Tórshavn, Sandur
Information Greenland Tourism a/s, Postbox 1552, DK-3900 Nuuk, Greenland, Ph. 299 22888; North Atlantic Marine Activity Ltd., PO Box 1371, FR-110 Tórshavn, Faeroe Islands, Ph. 298 12499; Joan Petur Clementsen, FR-210 Sandur, Faeroe Islands, Ph. 298 86119

237

Britain and Ireland

Europe and Africa

Britain and Ireland serve as the summer feeding areas for some whales, and provide year-round habitat for various groups of resident dolphins. Cruising through island passages dotted with castles and crofts, boat tours meet playful minke whales and bottlenose dolphins—the two most commonly sighted species. With their tag-along friendliness, they are welcome companions. It is also possible to see white-beaked, common, Atlantic white-sided, and Risso's dolphins; harbor porpoises; long-finned pilot whales; and orcas in the waters surrounding Britain and Ireland.

Scotland is becoming one of the headquarters for small-cetacean watching in Europe. Educational study tours directed toward minke whales have become popular offshore from the Isle of Mull. Risso's dolphins, and sometimes harbor porpoises and orcas, can also be seen. Off northern Scotland, an eight-hour ferry crossing from South Ronaldsay, in the Orkney Islands, to the Shetland Islands, provides one of the better runs for regular cetacean sightings in the British Isles.

The Moray Firth of Scotland offers the most northerly resident population of bottlenose dolphins in the world, and can be accessed by rail and air through the city of Inverness. Here, the bottlenose dolphins can be seen from land, feeding and playing. The best time for viewing is two to three hours before high tide from North Kessock, near the site of a small marine research base that monitors the dolphins and local seals.

In England, bottlenose dolphins are found off Cornwall and appear to move regularly between there and Cardigan Bay, Wales. However, the dolphins are more reliably seen out of Wales, with almost year-round day tours offered from New Quay, in Dyfed. Sightings are also sometimes made of common and Risso's dolphins, pilot, and even minke whales on extended day trips.

In the green country of coastal southern Ireland, the cetacean walking tour with bed

Fungie, the famous Dingle dolphin, leaping for tourists watching from a boat off the Dingle Peninsula, County Kerry, Ireland.

The rugged coastline of the Orkney Islands, off northern Scotland (above). Risso's dolphins (right) are seen off Britain and Ireland, mainly in deep water.

and breakfast, and pub stops, is an Irish innovation. Some of the best tours are those to Mizen Head and Clear Island, in County Cork. From two vantage points, Blannarragaun and the Bill of Clear, 12 cetacean species have been spotted, including harbor porpoises; Risso's, bottlenose, common, Atlantic white-sided, and white-beaked dolphins; and minke and long-finned pilot whales. A good source for information on current sightings is the Cape Clear Observatory. There is a ferry service to the island and a variety of accommodations available. Boat tours are also in operation to Mizen Head and Clear Island, along with other offshore areas.

In western Ireland, in the Shannon River estuary, some 60 resident bottlenose dolphins can be seen on year-round boat tours from Carrigaholt.

SPECIAL FEATURES

Communities in both Britain and Ireland have long had relationships with solitary friendly bottlenose dolphins. Dolphins named Freddy, Donald, Percy, and Simo were some of those who befriended people and took up residence in British coastal waters for two to three years. The latest dolphin is Fungie, who moved into Dingle Harbour, County Kerry, Ireland, in the mid–1980s, and began approaching and following boats. Leaping and spyhopping, he has entertained more than a million visitors since then, at the rate of up to 150,000 people a year. There is no way to predict how much longer he will stay— by 1996, he had remained longer than most others. Fungie has already done his bit to get many people interested in cetaceans. In future there will no doubt be other solitary dolphins seeking human company.

TRAVELER'S NOTES

When to visit Year-round dolphins but best seen May–Oct; Apr–Oct minke whales in western and northern Scotland; June–Aug prime whale- and dolphin-watching season

Weather Cool to cold on the water; rain possible, even in summer, especially in western parts of Britain and in Ireland

Types of tours Half- and full-day tours, extended multi-day expeditions; inflatables, sailboats, and large whale-watch boats; some whale watching from ferries; land-based whale watching

Tours available Britain: (England) Cornwall; (Wales) New Quay, Milford Haven; (Scotland) Dervaig (on the Isle of Mull), Mallaig, Oban, Gairloch, Cromarty, Inverness; Republic of Ireland: Carrigaholt, Dingle, Schull, Castlehaven, Kilbrittain, Clifden

Information British Tourist Authority, Thames Tower, Blacks Road, Hammersmith, London W6 9EL, England, Ph. 44 181 846 9000; Irish Tourist Board, Baggot Street Bridge, Dublin 2, Ireland, Ph. 353 1 602 4000

239

The Mediterranean

Europe and Africa

The Mediterranean is the world's largest inland sea, covering an area of nearly 1 million square miles (2.6 million sq km). Lying at the crossroads of Africa, Asia, and Europe, this saltier-than-usual sea connects to the open North Atlantic through the narrow 9 mile (14.5 km) wide Straits of Gibraltar, and to the Red Sea and Indian Ocean through the Suez Canal.

Sperm whales (above) are among several species found off Greece in summer.

Few people who visit Italy, Monaco, or the south of France on holiday realize that fin whales spend their summers just offshore. Their preferred feeding habitat is a part of the Mediterranean west of northern Italy called the Ligurian Sea. Over the past few years, whale watching here has grown steadily in popularity, with departures mainly from San Remo and Imperia in Italy, and Toulon and other ports in southern France. During winter, fin whales are thought to undertake short migrations within the Mediterranean, probably wintering mainly off the adjacent North African coast near Tunisia.

Elsewhere in the Mediterranean, particularly east of southern Italy, off Greece, there is a chance to see sperm whales and Cuvier's beaked whales. The water off southern Greece is up to 3½ miles (5.5 km) deep, more than enough for squid and other prey preferred by these two deep-sea hunters. Elsewhere in the Mediterranean are numerous bottlenose, common, striped, and other dolphins—playing, feeding, and raising young year-round.

The Mediterranean offers a wide variety of whale-watch tours. There are summer day trips to the Ligurian Sea from Porto Sole, near Genoa, Italy, aboard a large passenger sailboat. Field-study tours, in which participants help scientists for a week or more, are also available. As well as fin whale watching in the Ligurian Sea during the summer months, these field-study tours feature common and bottlenose dolphins in the Ionian Sea of Greece, and bottlenose dolphins around the islands of Losinj and Cres in Croatia.

There are also three- to eight-day summer sailing cruises in which you can join a "stage," or part of the trip. Cetaceans are not seen every day, but are a part of the overall trip; most participants will see dolphins coming to the bow,

Striped dolphins (above) and other species are seen year-round. The Italian city of Liguria (right), on the Mediterranean coast.

as well as fin whales, and potentially even more. Departures are from Toulon, France, and Almería, and sometimes Barcelona, Spain. Almost all long-range Mediterranean trips are sold through Tethys, a cetacean research group in Italy, or Wolftrail, a field-study tour agency in the Netherlands.

From Sheppards Marina, Gibraltar, tour operators take visitors to see three main species found year-round in abundance: bottlenose, striped, and common dolphins. Orcas sometimes appear in May and June. Whales, including sperm, pilot, and others, are found in Gibraltar Bay, as well as in the Straits of Gibraltar, but the dolphin sightings are much more reliable.

SPECIAL FEATURES

The Ligurian Sea may soon become the first marine-protected area for cetaceans in the Mediterranean. In 1992, the governments of Italy, Monaco, and France designated a marine sanctuary to protect Mediterranean cetaceans. Although final terms have yet to be agreed on, researchers and whale watchers in this important feeding area for whales, dolphins, birds, and fish are already working to help protect the area. Habitat protection is crucial for cetaceans and their dependent life systems in the Mediterranean, which carries one of the world's oldest transportation networks and is subject to intensive agricultural, industrial, and residential waste. Pollution is a growing, persistent problem and as cetaceans are one of the key indicators of the health of the environment, making a place for them is vital. Yet habitat protection will not be enough; concerted efforts by some 20 countries bordering the sea will be needed if pollution standards are to be improved and the Mediterranean kept healthy.

TRAVELER'S NOTES

When to visit June–Sept best for most cetaceans, although they are present year-round; May–Oct best for dolphins around Gibraltar, but tours also Nov–Jan
Weather Sept hot, dry, although sea breezes keep temperatures comfortable; May–Oct can be cool with occasional rough seas; Nov–Apr cool, sometimes cold at sea with rainstorms
Types of tours Half- and full-day tours, extended multi-day expeditions; inflatables, sailboats, fishing boats
Tours available Italy: Porto Sole, San Remo, Imperia; France: Toulon; Gibraltar; Spain: Almería, Barcelona; Greece: Kalamos Island; Croatia: Veli Losinj
Information Tethys Research Institute, Acquario Civico-viale G.B Gadio2, I-20121 Milano, Italy, Ph. 39 2 7200 1947; Wolftrail, Steenbakkersweg 25, PO Box 800, 7550 Av. Hengelo, The Netherlands, Ph. 31 742 478 985

Canaries and Azores

Europe and Africa

The Canary Islands, just 60 miles (100 km) off the northwest coast of Africa, due west of southern Morocco, include seven main and several uninhabited islands. Tenerife, the base for most of the whale watching, has a pale red volcanic landscape, with the highest point at 12,198 feet (3,781 m). Tourism development has taken over parts of the island, but it is still possible to see the traditional cave-like houses carved into the mountains.

Many whale-watch boats depart from Los Cristianos and Colón, near Playa de las Americas, on the southern end of Tenerife, accessible through two international airports. Tenerife is one of the few places in the world to offer year-round whale watching and, since the mid-1990s, along with southern New England and Quebec, it has become one of the three most popular whale-watch areas in the world. Choose your trip carefully though, not just based on price, but by looking for boats that offer naturalist guides and those that contribute to local research and conservation.

The warm, deep waters around Tenerife support a resident population of at least 500 short-finned pilot whales. These whales, between the size of a bottlenose dolphin and an orca, and sharing behavioral habits of both, live in tight family groups. Their characteristic bulbous heads,

Common dolphins off Pico, Azores (above) and the pale red cliffs (below, far right) along the southern coast of Tenerife, the Canaries.

hook-like flippers, and long bodies can be seen in the clear waters as they swim just beneath the surface before diving deep to hunt squid. Bottlenose dolphins are often encountered, along with the pilot whales, within an hour of leaving port.

Whale and, especially, dolphin watching has recently spread to Gomera, Lanzarote, and Gran Canaria. Some of the other islands offer chances to see inshore bottlenose dolphins (Gomera), or to sail in search of sperm whales and various beaked whales (Gomera, Lanzarote).

In the Azores Islands, in the North Atlantic west of Portugal, whales have long been a topic of conversation. Once a whaling center, they now offer whale watching from May to October, especially from the islands of Pico and Faial.

Relics of the old whaling days, vigias, or watch towers, such as the one seen above on Faial, are now used for research. Part of a large resident population of short-finned pilot whales (left) off Tenerife, in the Canaries.

Whale watching in the Azores is small scale compared to the Canary Islands, but the diversity of whales is greater. Sperm whales are a reliable sight, along with bottlenose, common, Risso's, spotted, and striped dolphins, plus short-finned pilot whales and various beaked whales. Few places in the world are as good for sighting rare beaked whales. A wide range of other species can also be seen.

SPECIAL FEATURES

In the Azores, traditional small boat–based whaling with hand harpoons was carried on until 1985, with three additional whales taken in 1987. Now the "vigias," or lookout towers where the whalers kept watch for the whales, have been restored by whale-watch companies. Many of them, especially on Pico and Faial where whale-watch tours are offered, are open to visitors. You may meet researchers who do some of their tracking of the whales from vigias.

One of the best is Vigia da Queimada, which allows sightings for up to 18 miles (30 km) on a clear day, within a radius of 200 degrees from east to west. At this distance, sperm whales and many other deep-water cetaceans can be seen. In Lajes, on the island of Pico, you can take a marked three-hour walk designed for whale watchers. This includes the Vigia da Queimada, the whaling museum, the old whalers' boathouse, and the whaling factory.

TRAVELER'S NOTES

When to visit Year-round Canaries: short-finned pilot whales, bottlenose, common, and many other tropical dolphins, along with small, toothed whales. May–Oct Azores: sperm, other whales, but whales may also be seen before and after this period

Weather Canaries: subtropical year-round with cool, refreshing winds; sometimes hot, sandy desert winds from Africa make whale sighting difficult, but there are usually 300 good whale-watching days per year; May–Oct Azores: seas windy, cool to cold

Types of tours Half- and full-day tours, some extended multi-day expeditions; sailboats and large whale-watch boats.

Tours available Canary Islands: Los Cristianos and Puerto Colón, near Playa de las Américas, on Tenerife, Gomera, Lanzarote, Gran Canaria; Azores: Horta on Faial; Lajes on Pico

Information Servicio de Turismo, Plaza de España S/N, Bajos del Palacio Insular, 38003 Santa Cruz de Tenerife, Tenerife, Canary Islands, Spain, Ph. 34 22 239 592; Azores Tourism Office, Casa Do Elogio Colonia Alema 9900, Horta-Açores (Azores), Portugal, Ph. 351 922 3801

South Africa and Madagascar

Europe and Africa

The Republic of South Africa has developed land–based whale watching to a fine art. Where the warm waters of the South Atlantic sweep around to meet the Indian Ocean, the main season for southern right whales is July to November. The whales come in close to shore then, to mate and raise their calves. Best access is through Cape Town, where you can rent a car and drive along the "Cape Whale Route." Along this ocean-front road, which extends from Cape Town east along the coast, there are dozens of lookouts along the rocky coastline that provide superb right whale watching. The whales come so close to shore that they are seen even from hotel rooms and local restaurants.

Hermanus, some 60 miles (95 km) east of Cape Town, is the center of activity, attracting tens of thousands of visitors from around the world. It has a festival atmosphere through the whale season; a "whale crier" strolling through the town blows a bass kelp horn to alert locals and visitors to sightings of right whales. He wears a small billboard showing locations of the sightings. Anyone interested is given a map and

Shore-based whale watching in Hermanus, South Africa, is a learning experience (above).

directions. The sounds on the horn also employ a sort of whale Morse code to indicate current sighting locations that change from day to day, sometimes from hour to hour. Special naturalist-led walking tours are offered by various tour companies in the town. A naturalist takes you to the best sites along the rocky cliffs and talks about the whales as you watch and wait.

At present there are no boat-based whale-watch tours to see right whales in South Africa, but it isn't necessary as they come so close to shore. Bottlenose and Indo-Pacific hump-backed dolphins can also be seen from shore in many

Bottlenose dolphins (above) are seen year-round in Madagascar and South Africa. Magellanic penguins (right) in Cape Province.

places along the coast, such as Plettenberg Bay. In the west of the country, out of Lambert's Bay, there are boat tours to see Heaviside's dolphin, found only in southwestern Africa. There are also a few boat tours on which humpback and Bryde's whales are seen.

Across Mozambique Channel, northeast of South Africa, in the Indian Ocean, is the island of Madagascar. The world's fourth largest island, at 229,000 square miles (593,000 sq km), it is almost the size of California and Oregon put together. Although more and more of its tropical forest has been cut down in recent years, this large subtropical island offers many natural wonders, and sightings of various dolphin species.

On Madagascar's east coast, around the island of Nosy Boraha (also known as Sainte-Marie),

humpback whales return every year from July to September to mate and raise their young. In Andampanangoy, boat tours can be booked through local hotels and outdoor sports shops. In addition to humpback whales, sperm whales can be spotted farther offshore, while inshore to offshore, bottlenose, spinner, and Indo-Pacific hump-backed dolphins are commonly found throughout the year.

SPECIAL FEATURES

With many parks close at hand in South Africa and Madagascar, visitors will want to experience rain forest, savannah, and the vast coast. Rare primates, such as lemurs, and other endemic flora and fauna are attractions on Madagascar. The Masoala Peninsula, adjacent to the prime humpback whale area, has pristine lowland rain forest where red-ruffed lemurs live.

A South African wildlife area close to the whale-watching sites is the Cape of Good Hope Nature Reserve. There are also a number of easily accessible colonies of Magellanic penguins well worth visiting in the Cape Province region.

TRAVELER'S NOTES

When to visit July–Nov southern right whales in South Africa; July–Sept humpback whales; and a chance to see mating behavior and calves in Madagascar; year-round bottlenose, Heaviside's, and Indo-Pacific hump-backed dolphins

Weather July–Nov in South Africa weather mixed with rain and strong winds alternating with warm-to-hot temperatures; July–Sept, humpback season in Madagascar; warm weather even at sea

Types of tours Mainly land-based lookouts and tours; boat tours for dolphin watching in South Africa; small-boat and fishing-boat tours in Madagascar

Tours available South Africa: Hermanus, Plettenberg Bay (various), Lambert's Bay; Madagascar: Andampanangoy

Information South African Tourism Board (SATOUR), Private Bag X164, Pretoria 0001, Republic of South Africa, Ph. 27 123 470600

245

ASIA

As Japan leads the way, whale watching, and especially dolphin watching, is catching on in fishing villages and beach resorts all over Asia.

INDIAN OCEA

Whale watching is proving its worth as a source of income in Asia, a vast region with important fishing economies where many people depend on the sea for their livelihood.

In April 1988, the first Japanese whale-watching expedition departed from Tokyo for the Ogasawara Islands. This trip, reported worldwide, was so notable because Japan remains both a whaling nation and a consumer of whale meat. Since 1988, Japanese whale watching has flourished. Some 25 communities around the country now offer whale- and dolphin-watching tours. With a whale-watching season of only three months a year, the Ogasawara whale-watch industry provides half of the year's tourist income to the islands. That is twice the revenue from agriculture and almost half that of the most important industry—fishing.

In some communities, the local government or fishing collective sponsors whale watching. In other areas, local fishers, dive operators, and even some professional people

have organized the whale-watch tours. Many Japanese travelers have experienced whale watching in Hawaii, Canada, or Baja California, so it was perhaps inevitable that they would take ideas back to their own country. The surprise is how popular it has become. More than 60,000 people a year—most of them

Japanese—now take whale- and dolphin-watching tours in Japanese waters.

In Hong Kong, dolphin tours have already become popular. The tours are based

Searching for whales near Krakatau volcano (above), in Indonesia. A pod of dwarf sperm whales (below left).

on the resident pink Indo-Pacific hump-backed dolphins and finless porpoises that are seen in some of the most well-traveled waterways of the world. In Hong Kong, dolphin watching has alerted many people to the problems of conserving dolphins, which

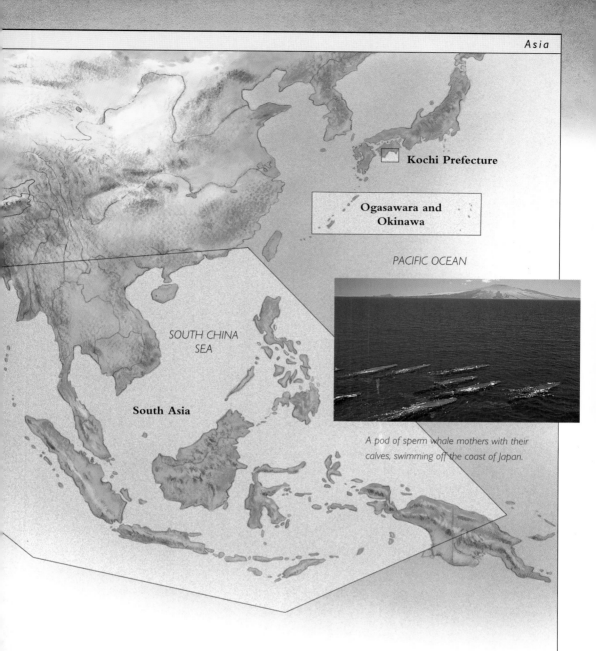

Kochi Prefecture

Ogasawara and Okinawa

PACIFIC OCEAN

SOUTH CHINA SEA

South Asia

A pod of sperm whale mothers with their calves, swimming off the coast of Japan.

are being displaced from local waters by industry. In Japan and China, along with other countries in Asia, the hope is that cetacean watching will expand and become even more important as an economic, educational, scientific, and conservation force. If this happens, there's a better chance that cetacean populations will increase and flourish.

In other parts of Asia, whale watching is newer and still developing. In Goa, India, small boats take holiday beachgoers to meet tropical dolphins just offshore. The same is true in Bali, Indonesia;

off Phuket, Thailand; and off the east coast of Taiwan. Because of the great diversity of species offered, from sperm whales to rare dolphins, the Philippines has the potential to appeal to discerning and dedicated whale watchers and ecotourists. Educational and

Whale watching aboard a Japanese fishing boat (above) in Ogata, Japan.

conservation programs have been established, and the Philippines may one day challenge even Japan for the title of hottest whale-watching ticket in Asia.

247

South Asia

Asia

Even more than in other parts of the world, tropical cetacean watching in South Asia can be extraordinarily diverse. Just a short boat ride from Queen's or Kowloon Piers, in Hong Kong, are wild dolphins, which is one of the last things you would imagine seeing in that environment. Close to Hong Kong's new airport, a dozen pinkish dolphins lift their backs out of the water to display the characteristic hump of their species. These are Indo-Pacific hump-backed dolphins—a resident population found in the inshore waters of eastern China.

According to photo-identification studies, about 250 hump-backed dolphins use the North Lantau area of Hong Kong, living there year-round, feeding on a wide variety of fish—including gray mullet, white herring, shad, pilchards, and anchovies, as well as shrimp.

Along with the hump-backed dolphins are tiny finless porpoises. Looking like mini belugas, they butt the water with their round foreheads, poking around in the shallows. With neither a beak nor a dorsal fin, they are easy to distinguish

Tiny finless porpoises (left) near Hong Kong. Wooden boats on the beach in Goa, India (below right).

from the hump-backed dolphin, with its long beak and prominent dorsal fin.

In the Tañon Strait of the Philippines, a wide variety of tropical whales and dolphins can be encountered. Dwarf sperm whales, a rare species anywhere else, can be seen swimming in family groups. Also watch for spotted, spinner, Risso's, and Fraser's dolphins, as well as pilot and melon-headed whales. Sperm and Bryde's whales are sometimes seen. Dolphins appear year-round, but the best months for spotting many species are April to June.

In Indonesia, sperm whales reside off Lembata Island in East Nusa Tenggara and off eastern Java, although tours are only occasionally available. From Bali, however, there are popular organized tours from Lovina Beach on the north side of the island where gregarious, acrobatic common and spinner dolphins can be seen.

In Goa, on the west coast of India, bottlenose and striped dolphins can be encountered just offshore. The small, wooden boats found along

Hong Kong's busy harbor (above). Melon-headed whales in Tañon Strait (left).

the beach offer impromptu tours of several hours. Thailand also offers dolphin watching as part of diving, marine-nature, and sea-canoe adventure trips out of Phuket. Dolphins are also sometimes advertised as part of trips to the Similan and Surin Islands in the Andaman Sea, but Thailand's potential is still largely undeveloped.

Most smaller cetaceans feed, socialize, and mate in the waters of South Asia year-round. Only baleen whales migrate to cooler waters to feed.

South Asia is one of the best areas to visit if you want to help save whales and dolphins. The survival of local whale and dolphin populations depends partly on the ability of local communities to manage cetacean tourism—to make a living from showing people "their" local wildlife.

SPECIAL FEATURES

In the early 1980s, a three-year World Wide Fund for Nature (WWF) research expedition in the Indian Ocean identified a hotspot for whales and dolphins around Sri Lanka, south of India. Year-round populations of spinner and bottlenose dolphins were found, and there were sporadic sightings of striped, spotted, Risso's, Irrawaddy, and Indo-Pacific hump-backed dolphins. Most impressive of all are the large whales, including blue, sperm, and Bryde's whales, although the blue whales are present only from February to April.

Oceanic Society Expeditions in San Francisco was the first to offer trips in the 1980s. But the civil war in Sri Lanka forced cancellation of the program. Visitors to Trincomalee may be able to go a mile northeast to a military site called Fort Frederick and scan the Bay of Bengal. Within an hour, several cetacean species can often be spotted.

TRAVELER'S NOTES

When to visit Year-round, dolphins throughout region, sperm whales off Sri Lanka; Feb–Apr, blue whales off Sri Lanka; Apr–June, best for Philippines
Weather Hot, even at sea, but weather conditions vary across this vast region; main obstacles are prevailing winds that make whale watching in exposed areas difficult when the surf starts to build
Types of tours Half- and full-day tours, some extended multi-day expeditions; small boats, sailboats, canoes, outrigger boats, and motor cruisers;

some land-based whale watching. Visitors arriving during storm and monsoon seasons should pay attention to local weather advisories
Tours available Hong Kong; India: Goa; Thailand: Phuket; Indonesia: Lovina Beach on Bali; Philippines: Tagbilaran on Bohol
Information Hong Kong Dolphinwatch, GPO Box 4102, Hong Kong, China, Ph. 852 2984 1414; Tourist Information Center, Jalan, Legian, Kuta, Denpasar Bali, Indonesia, Ph. 62 361 751551 or 62 361 751875

Japan: Ogasawara and Okinawa

Asia

In two tropical Japanese archipelagoes—the Kerama Islands off Okinawa and the Ogasawara (or Bonin) Islands—humpback whales visit every winter to sing, mate, and raise their calves. The Ogasawara Islands, 600 miles (1,000 km) south of Tokyo, are part of a volcanic offshore chain. At the southern end, Chichi-jima and Haha-jima, two inhabited islands, have long been popular with divers, underwater photographers, and nature tourists. Whale watching began here in 1988. Access is by a 29-hour ferry ride from Tokyo.

A typical whale-watch day in Chichi-jima begins with a trip up the hill for a talk on whales and a chance to scan for cetaceans from land. Often, long humpback flippers can be seen waving or slapping the surface. Until a few decades ago, from various Japanese ports, whales were spotted from strategic lookouts on land, alerting the whalers to leave port. These days scientists from the Ogasawara Whale-watching Association guide visitors and report sightings to fishing boat captains who take tourists out to find

The humpback (left) is just one of a number of cetacean species seen off the Kerama and Ogasawara islands.

the whales. Besides humpbacks, other species commonly seen around Ogasawara are Bryde's, sperm, and short-finned pilot whales, along with bottlenose and spinner dolphins. There are also common, Pacific white-sided, spotted, rough-toothed, Risso's, striped, and Fraser's dolphins. Dolphin watching in this area is a popular, but separate, pursuit to whale watching, with operators specializing in watching or swimming with (mainly) bottlenose dolphins.

Access to the Kerama Islands is by jet airplane through Naha on Okinawa, then a short ferry ride to Zamami (the main whale-watch village) or Tokashiki. The Keramas are famous for diving opportunities, and many whale-watch operators in Zamami offer diving tours throughout most of the year, although some tours concentrate only on whales during the whale-watch season. The prime areas for diving are the relatively untouched coral reefs around the islands. A few

The coral seacoast (above) off Okinawa. A pod of spinner dolphins (left) leaping in the waters off the coast of Ogasawara Island.

SPECIAL FEATURES

Japanese scientists aboard whale-watch boats in Ogasawara and the Keramas have compiled a photo-identification catalog of more than 600 individual humpback whales. A female humpback whale photographed in April 1990 and March 1991 in Ogasawara, turned up in August 1991 on the summer feeding grounds off British Columbia, some 4,900 miles (7,900 km) away. Those feeding grounds are commonly used by humpback whales from Hawaii, but not Japan. Yet a few humpbacks are known to migrate to British Columbia not only from Japan and Hawaii, but from Mexico as well. This indicates some exchange on the feeding grounds among the three main breeding areas (Hawaii, Mexico, and Japan) for North Pacific humpback whales.

of the local operators offer visitors the chance to dive in the morning and whale watch in the afternoon, or vice versa.

In the Kerama Islands, humpback whales, bottlenose dolphins, and the comparatively rare rough-toothed dolphins are commonly seen. Occasional sightings are made of false killer and short-finned pilot whales, along with common, spinner, and spotted dolphins.

Whale-watch clubs are popular in Japan. At both Ogasawara and Zamami, the local whale-watching associations send out newsletters, design T-shirts for sale, and manufacture buttons and bumper stickers. They also help manage the promotion and sales of whale-watching trips. These are two of the most active and popular whale-watch clubs in Japan. Each has hundreds of members from all over the country and abroad. Club members regularly receive news of their favorite whales, whale-watch schedules, invitations to special events, and special prices for whale watching.

TRAVELER'S NOTES

When to visit Feb–Apr, humpback whales winter on the mating and calving grounds around Ogasawara and Okinawa and the Kerama Islands; year-round, various dolphins

Weather Warm to hot during the season, but often windy and cool at sea

Types of tours Half-day tours; inflatables, diving and fishing boats; some whale watching from ferries

Tours available Chichi-jima and Haha-jima in the Ogasawara Islands; Zamami and Tokashiki in the Kerama Islands; Naha on Okinawa

Information Ogasawara Tourism Association, Chichi-jima, Ogasawara-mura, Tokyo 100-21, Japan, Ph. 81 4998 2 2587; Zamami Village Office, 109 Zamami, Zamamison-aza, Okinawa-ken, 901-34, Japan, Ph. 81 98 987 2311

Japan: Kochi Prefecture

Asia

Whale watching Japanese-style has a particularly distinctive flavor. In the town of Ogata, the most popular whale-watching community in Japan, there are Shinto religious ceremonies to welcome the whales back every year and to pray that they will always return. The overlay of Japanese culture also creates some unusual whale-watch "artifacts," from handmade, individually wrapped whale cookies and whale-brand pickled onions, to the familiar array of T-shirts and buttons with Japanese logos and English words often used as design elements on them. Ogata is 2½ hours by train away from Kochi City, the capital of Kochi Prefecture, which is accessible by rail or plane from Osaka or Tokyo.

The whales watched at Ogata are mainly Bryde's whales. Photo-identification studies show that at least 15 Bryde's whales are resident in Tosa Bay. From March to October, they come in close to shore, feeding in the shallows near the town. In most cases, the whales can be found within an hour of leaving Ogata. These whales sometimes breach, sliding their long,

slim bodies out of the water and high up into the air, showing off their white or pinkish throat grooves. One Bryde's whale set a local record by breaching 70 times in a row, to the delight of whale watchers. In winter, the whales remain offshore, venturing close less often.

Whale watching in Ogata began in 1989, a year after it started in Ogasawara (see p. 250), but Ogata has since taken the lead, attracting three or four times as many visitors. Several communities north and south along the coast from Ogata, including Saga, Shimonokae, and Tosa-shimizu, have also taken up whale watching. While boats sometimes encounter various dolphins, along with orcas and short-finned pilot whales, the species most commonly sighted is Bryde's whales.

On the eastern side of Tosa Bay, some six hours away by car but still in Kochi Prefecture, is the city of Muroto. Here, one of the best whale-watch opportunities in the world is led by a former whaler, now converted to whale

Risso's dolphin (left) breaching off Cape Muroto. A former whaler (right) turned whale watcher.

252

The coast along Cape Ashizuri National Park (above) is great for whale watching. Close up of Bryde's whale (left), Ogata.

watching. Throughout most of the year, except January, a converted fishing boat cruises the deep waters just off Cape Muroto where whales were once approached only by harpoon. There are also occasional ferry excursions sponsored by the city. Sperm whales are the main quarry, followed by Risso's and bottlenose dolphins and short-finned pilot whales. At least 10 other species of whale and dolphin are also occasionally seen.

SPECIAL FEATURES
During the whale-watch season, special events in Ogata include the annual summer T-shirt festival and a sandcastle contest. Handmade and

beautifully designed T-shirts from all over Japan are mounted on sticks and exhibited on the beach. Winning sandcastles have included full-sized whales and medieval castles in extraordinary detail.

The pure waters of the Shimanto River make their way through the mountains of Kochi Prefecture down to the beach near Ogata, close to where the Bryde's whales come in to feed. Near the river, at Nakamura City, is the famous Dragonfly Museum and Nature Reserve. Along the coast are steep, heavily forested cliffs in Cape Ashizuri National Park with advantageous whale lookouts beside the road. At Tosa-shimizu, the John Mung House commemorates a Japanese pioneer who has been credited with opening doors to the West and, among other things, bringing modern whaling to Japan.

TRAVELER'S NOTES

When to visit *Mar–Oct (best May–Sept), Bryde's whales off Ogata, Saga, and other communities in Kochi Prefecture (some sightings through other months, but infrequent); year-round, dolphins in most locales, plus short-finned pilot whales and sperm whales off Cape Muroto, near Muroto, but best Mar–Dec*
Weather *Oct, varies from cool to warm at sea, depending on local currents and weather patterns; Nov–Feb, cool to cold at sea; typhoon season Aug to early Oct*

Types of tours *Half- and full-day tours; fishing boats; some watching from ferries*
Tours available *Ogata, Saga, Shimonokae, Tosa-shimizu, Kochi City, Cape Muroto*
Information *Fisheries and Commerce Section, 2019, Irino, Ogata-cho, Hata-gun, Kochi-ken, 789-19, Japan, Ph. 81 880 43 1058; Muroto City Office, Commerce & Tourism Section, 25-1, Ukitsu, Muroto-shi, Kochi-ken, 781-71, Japan, Ph. 81 887 22 1111*

AUSTRALIA
and OCEANIA

Whale watchers in Australia and Oceania see resident Southern Hemisphere cetaceans, as well as those using the waters for migration, mating, and calving.

Whale watching throughout the waters of Australia and Oceania has mostly developed during the 1990s. Australia and New Zealand have led the way, aided by well-developed domestic travel industries that help support whale watching, as well as attracting visitors from around the world.

In Australia, humpback and southern right whale watching was mostly land-based until 1987, when boat-based tours operating out of Hervey Bay, Queensland, gained in popularity. Along with members of many communities near the eastern and southwestern coasts of Australia, watchers avidly follow the annual migrations of the humpbacks.

Australia also offers outstanding destinations for viewing migrating southern right whales. In Warrnambool, Victoria, since the early 1980s, the southern right whale nursery at Logan's Beach has attracted thousands of visitors every year. Some adventurous tourists trek to the head of the Great Australian Bight or the Bunda Cliffs, along the Eyre Highway, for some of the most spectacular cliffside whale watching in the world.

East Coast Australia

INDIAN OCEAN

Western Australia

Southern Australia

SOUTHERN OCEAN

Coastal cliffs (top) of South Australia. A southern right whale and dolphins (left) off Australia's south coast.

At Monkey Mia, in Western Australia, bottlenose dolphins have enjoyed a friendly relationship with humans for many years. Thousands of people visit every year, wading in the shallow water where dolphins come to interact with their land-based neighbors. Several other dolphin populations are found in inshore waters around Australia and these have awakened the keen interest of researchers and whale watchers alike.

New Zealand's whale watching began around the same time as Australia's, but has since taken off in different directions and with a flavor all its own. Whales spotted around New Zealand are largely toothed whale species that are resident rather than migratory. Sperm whale watching at Kaikoura, New Zealand's premier watching site, has helped create one of the world's most charming communities dedicated to whale watching. Elsewhere around New Zealand, rare dolphins lend a special magic to the watching—the Hector's dolphin is found only in New Zealand waters. Dusky dolphins, also seen in a number of other Southern Ocean locations, are still most frequently spotted here.

Although the whale-watching opportunities in the South Pacific islands of Oceania are less developed than those in Australia and New Zealand, nearly every year another island sets up watching opportunities.

These are boosting tourism, which means that the cetaceans may be better looked after. Such tours are sometimes associated with research work or, occasionally, diving tours. Tonga has taken the lead with humpback tours to meet mothers and calves wintering near the islands, but other Pacific islands also have nascent dolphin-watching industries. These include Tahiti, Niue, Moorea, Fiji, New Caledonia, and Western Samoa.

Humpback mother and calf, Tonga, South Pacific Ocean.

Vanuatu
Fiji
Western Samoa
Tahiti and Moorea
Tonga and Other Pacific Islands
Tonga
Niue
New Caledonia

PACIFIC OCEAN

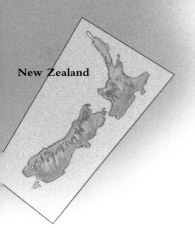

New Zealand

Dusky dolphins playing off Kaikoura, New Zealand's premier watching site.

East Coast Australia

Australia and Oceania

Huge humpbacks roll over and over in the wake of the boat, as if to show off the extensive whiteness of their bellies, flippers, and flukes. Their immense bodies are set off by cascades of froth and foam. It's morning in Hervey Bay, and the sun is rising over nearby Fraser Island, the world's largest sand island. Hervey Bay, the whale-watch capital of Australia, is 180 miles (290 km) north from Brisbane.

Between August and October, several hundred migrating humpback whales pass through the sandy strait between Fraser Island and the mainland. The first to arrive, from their winter grounds around the Whitsunday Islands, are the older juveniles, followed by the mature males and, finally, mothers and calves. These whales stop off in the warm, sheltered waters of Hervey Bay before moving on to their final summer destination, Antarctica, where they feed on fish and krill.

The humpbacks of the Southern Hemisphere are known to be whiter on the underside than their northern counterparts, although both are considered the same species. In recent years, an all-white humpback—an albino—has been sighted off eastern Australia.

Most of the whale-watch boats depart from Urangan Harbour, Hervey Bay, but others leave from Bundaberg, farther north, and other towns along the Queensland coast. The humpbacks, as well as various dolphins, can be seen in the waters between the Great Barrier Reef and the mainland. From Airlie Beach, tours depart to see humpback whales on their winter breeding grounds in the waters surrounding the Whitsunday Islands.

In New South Wales, south of Queensland, humpbacks pass Byron Bay at Cape Byron, Australia's most easterly point. Land-based whale watchers have assisted researchers here every year by counting the humpbacks passing below the cliffs. These whales come closest to shore and in the greatest numbers during June and July.

A humpback whale shows its pale ventral side as it swims upside down in Hervey Bay off the east coast of Australia.

256

Seven bottlenose dolphins surfing off the coast south of Wollongong (above). Cape Byron (left), Australia's most easterly point.

Some 50 miles (80 km) south of Sydney, at Wollongong, humpback tours have a split season. For the northward migration, the season is late May to mid-July; on the southward journey, the humpbacks pass from late September to November. Dolphin sightings include mainly bottlenose, as well as common and Risso's dolphins.

The town of Eden, about 300 miles (485 km) south of Sydney, on Twofold Bay, offers a combination of historical whaling interest and whale-watch tours. These tours can catch migrating humpbacks during their northerly migration in June and July and their return south in October to November. Because Twofold Bay is near the far southeastern corner of Australia, it is both the first and last land point the humpbacks pass on

their journey to and from Antarctica. It is also possible to see bottlenose dolphins here and, sometimes, southern right and blue whales, as well as Australian fur seals.

SPECIAL FEATURES

Visitors to Hervey Bay may want to spend an extra day or more exploring sandy Fraser Island, listed as a World Heritage site. It has tropical rain forest, heathland covered in wildflowers, and long, sandy beaches. Most of it is protected as a national park. Seventy miles (112 km) long and 14 miles (22 km) across at its widest point, Fraser Island has numerous lakes, hiking trails, and hundreds of native animals and birds. The forest trees include swamp mahoganies, forest red gums, and various eucalypts, including blackbutts and scribbly gums. Don't miss Rainbow Gorge and Woongoolver Creek, the Maneno shipwreck, the Cathedral Sandcliffs, and the wild horses. Along the high dunes on the west side of the island are excellent land-based lookouts for humpbacks, often seen playing close to the island along Platypus Bay. Remember to take your own food and water.

TRAVELER'S NOTES

When to visit *Humpback whales: Queensland, including Hervey Bay, Aug–Nov; New South Wales, including Wollongong, late May–mid-July, late Sept–Nov, and Eden, June–July, Oct–Nov. Dolphins: year-round*
Weather *Queensland coast, warm to hot on humpback breeding grounds; New South Wales, cool to cold on water*
Types of tours *Half- and full-day tours, some extended multi-day expeditions; inflatables, sailboats, diving boats, and large whale-watch boats;*
land-based whale watching also good
Tours available *Queensland: Airlie Beach, Bundaberg, Hervey Bay, Tangalooma; New South Wales: Byron Bay, Coff's Harbour, Eden, Fairy Meadow, Wollongong*
Information *Queensland Tourist and Travel Corp., Level 36, Riverside Centre, 123 Eagle St., Brisbane, Queensland 4000, Australia, Ph. 61 7 3406 5400; Tourism New South Wales Travel Centre, 19 Castlereagh St., Sydney, New South Wales 2000, Australia, Ph. 61 2 132 077*

Southern Australia

Australia and Oceania

Southern Australia has some of the most outstanding land-based whale-watch sites in the world. Photographs taken from the cliffs and beaches are the equal of close-up photographs from boats—with the added advantage of a firm shooting platform. Forget about being seasick.

In South Australia, endangered southern right whales come in close to shore. Tours are rugged, multi-day excursions across the outback—a chance to see a wide variety of South Australia's native wildlife, as well as the spectacular coast around the Great Australian Bight.

Southern right whales can be seen all along the Bunda Cliffs, tracing an 80 mile (130 km) stretch of the Eyre Highway through Nullarbor National Park. Visitors can stay at the caravan park, the Nullarbor Hotel-Motel, or camp in the national park. For a camping permit, contact the National Parks and Wildlife Service in Ceduna.

The most reliable spot for whale watching is farther east, at Twin Rocks, at the head of the Great Australian Bight. Up to 70 whales have

A pair of orcas spyhopping (left) off the coast of Southern Australia. Two common dolphins riding a wave at sunset (bottom right).

been seen at one time, some within 325 feet (100 m) of the shore. Twin Rocks is situated on Yalata Aboriginal land, 685 miles (1,100 km) by road from Adelaide along the Eyre Highway. All whale watchers must obtain permits at the Yalata or Nullarbor Roadhouses on the Eyre Highway, before attempting to cross Yalata lands.

More accessible to Adelaide, within an hour or two's drive, are various other southern right whale lookouts. In recent years, the most popular has been the Victor Harbor coastline, south of Adelaide, from the Bluff to Middleton Bay. Other possible sites include Waitpinga, Parsons Beach, and Goolwa.

Although the southern right whale is by far the most commonly seen whale, whale watchers should watch out for other cetacean species. At least 25 species have been recorded along Southern Australia, many of them from strandings, so beachcombers may come across

A southern right whale with her calf (above) off the coast from the Nullarbor National Park, South Australia. There is also excellent land-based whale watching (right) at Warrnambool, Victoria.

some as well, while looking for cetaceans surfacing and spouting offshore. Possibilities include orcas; bottlenose and common dolphins; humpback, blue, minke, pilot, false killer, and sperm whales; as well as various beaked whale species. This is one of the few places in the world where regular strandings of rare pygmy right whales occur.

Every year between May and late October, the city of Warrnambool, Victoria, attracts thousands of visitors who come to see the southern right whale nursery at Logan's Beach. These whales come in close to shore and can be spotted from various observation platforms. Logan's Beach is only 2½ miles (4 km) from Warrnambool, and 165 miles (265 km) from Melbourne.

SPECIAL FEATURES
Visitors to Australia's southeastern coast can watch dolphins, although this site is not as

popular as the world famous dolphin watching offered at Monkey Mia, in Western Australia (see p. 260). One accessible dolphin population in Victoria lives in Port Phillip Bay, which is one of Melbourne's busy holiday regions. More than 100 bottlenose dolphins living here have been photo-identified and studied since 1990 by the Dolphin Research Project, Inc. Like bottlenose dolphins everywhere, those found in Port Phillip Bay are typically friendly and curious. Boat tours to see the dolphins are also offered, especially around Point Nepean, and some years humpback, southern right, killer, and pilot whales, as well as orcas, come into the bay.

TRAVELER'S NOTES

When to visit *Year-round bottlenose dolphins at the Head of Bight, South Australia, and Port Phillip Bay, Victoria (summer best at south end of bay); May–Oct southern right whales at the Head of Bight and other bays along the South Australia coast; mid-June–Oct southern right whales at Head of Bight and Victor Harbor, SA, and Logan's Beach, Vic.*
Weather *Cool to cold on the water, but can be warm from sheltered lookouts*
Types of tours *Largely land-based whale watching, some organized as*

multi-day expeditions, but most are informal day trips; some half- and full-day boat tours, inflatables, and small boats out of Port Phillip Bay
Tours available *South Australia: Ceduna, Victor Harbor; Victoria: Moorabbin, Logan's Beach*
Information *South Australian Travel Centre, 1 King William St., Adelaide, South Australia 5000, Australia, Ph. 61 8 212 1505; Tourism Victoria, 55 Swanston St., GPO Box 2219T, Melbourne, Victoria 3001, Australia, Ph. 61 3 653 9777*

Western Australia

Australia and Oceania

Since the 1960s, the dream of meeting dolphins in the shallows of a sandy beach has attracted people from around the world to Monkey Mia. Located at the southern end of Shark Bay, on the Indian Ocean, Monkey Mia is 505 miles (810 km) north of Perth, and is the most frequented cetacean-watching site in Western Australia, with 100,000 visitors a year.

Visitors commonly wade into the warm, shallow water while the dolphins approach, swimming up to nudge bathers' legs, and allowing themselves to be touched. Those swimming close to shore are a small group of six bottlenose dolphins—currently three adult females and their three calves. From time to time, the individual dolphins have changed, yet the number has stayed at about six, and their interaction with humans continues, most days, year after year.

Local rangers are on hand to supervise, provide information, and make sure that only a few people enter the water at a time. Visitors stand in the shallows and watch the dolphins swim up to them. In the distance, and from other points along the shore, many more dolphins belonging to the population living in the area can be seen.

Because of the accessibility of these dolphins and the clear-water viewing conditions, world researchers have taken the opportunity to study their behavior. Much of what we know about

A ranger keeps watch on proceedings as a bottlenose dolphin (above) greets tourists in the waters of Monkey Mia (right).

bottlenose dolphin social behavior comes from those studied at Monkey Mia, as well as another bottlenose group resident in the waters off Sarasota, Florida, in the United States.

Visitors to Monkey Mia during the humpback whale migration, from July to September, can also take boat tours or watch whales from land. Whale-watch boats depart from Denham, 15½ miles (25 km) away, or from Carnarvon, 217 miles (350 km) away. The best land-based spot is Point Quobba, north of Carnarvon. You can drive or fly to Denham and Carnarvon from Perth.

Besides the Shark Bay area, Western Australia has thriving boat-based humpback whale

A southern right whale calf (above) learns survival skills in the shallow waters around Cape Arid National Park, Western Australia.

watching on the southwest coast, especially around Perth. One of the most popular ports is Hillary's Boat Harbour, a few miles north of Perth. Boats also leave from Fremantle, south of Perth. Most of the humpbacks come through between September and November, staying close to shore, with many traveling between Rottnest Island and the mainland—easily accessible to whale watchers on half-day trips. Tour boats show a 97 percent success ratio with an average of four whales seen per trip. For the most part, the boats don't venture far offshore. In recent years, particularly out of Geraldton, about 250 miles (400 km) north of Perth, the boat trips in search of humpback whales have occasionally found Bryde's, sei, and minke whales.

SPECIAL FEATURES

The Albany area, 255 miles (410 km) south of Perth, is accessible by direct flight or road

and provides another window on whales in Western Australia. Albany was the last whaling site in Australia. Since the Cheynes Beach Whaling Station closed in 1978, the whaling boat *Cheynes IV* has been on display. Also worth a visit is Albany Whale World, devoted to the region's whaling history. Fortunately, some whales still survive and on regular tours from August to November, you can meet endangered southern right whales. Many tours offer a full experience of the region with a visit to the old whaling station and sightseeing around King George Sound to see local seals, bottlenose dolphins, pelicans, and other birds.

TRAVELER'S NOTES

When to visit Year-round bottlenose dolphins at Monkey Mia (Apr–Oct, dolphins approach swimmers most often, with less frequent sightings Nov-Mar); May–Oct southern right whales (Aug–Nov is prime for Albany area); Sept–Nov humpbacks in Perth area (July–Sept in northern part of Western Australia)

Weather Generally cool on the sea for boat-based whale watching; shore-based dolphin and whale watching can be warm or even hot in summer months, particularly at Monkey Mia

Types of tours Half- and full-day tours, some extended multi-day expeditions; inflatables, sailboats, and large boats; land-based whale watching

Tours available Perth, South Perth, Hillary's Harbour, Fremantle, Geraldton, Exmouth, Carnarvon, Albany, Denham, Monkey Mia (land-based watching)

Information Western Australia Tourism Commission, Floors 5 & 6, 16 Saint George's Terrace, Perth 6000, Western Australia, Australia, Ph. 61 8 9220 1700; Shark Bay Tourist Bureau, 71 Knight Terrace, Denham, Western Australia 6537, Australia, Ph. 61 8 99481 253

New Zealand

Australia and Oceania

Few communities in the world have become so captivated and identified with whales and dolphins as Kaikoura, New Zealand. Since the late 1980s, Kaikoura—a small town at the foot of snow-capped peaks on a point jutting out into the sea, on the east coast of the South Island—has become an idyllic spot to spend a few days, go whale and dolphin watching, maybe see some seals, or take the time to go hiking.

Sperm whale watching in Kaikoura is a year-round phenomenon. The main operation is run as a community trust by the local Maori people. Their boats boast a 97 percent success rate for sightings between April and July and about 95 percent for the rest of the year. Of course, on some days it is not possible to go out because of bad weather or rough seas, but when you are sitting in a friendly local cafe reading whale brochures, it is comforting to know that the whales will still be out there when the weather, with any luck, clears in the next day or two.

New Zealand's Hector's dolphins (left), Banks Peninsula. Watching dusky dolphins (above right).

Along with sperm whale watching, Kaikoura offers specialized tours for dolphin watching, swimming with dolphins, and whale watching from a helicopter or seaplane. You can swim with pods of playful dusky dolphins in the morning and fly over them and the sperm whales by helicopter or seaplane in the afternoon. Don't worry: each type of whale watching has its own regulated code of contact, so the animals are not disturbed. Aircraft, for example, cannot swoop too low, as the sound might bother the whales.

There are also kayak tours and various land-based hiking and birding tours on offer. Best access to Kaikoura is through Christchurch airport followed by a few hours' drive north along the coast of the South Island.

The sperm whales living off the coast from Kaikoura are bachelor males (see also p. 95). Within the deep canyons located close to shore,

The snow-capped Kaikoura Range (above) provides a dramatic backdrop to New Zealand's most popular whale-watching area.

they hunt and catch squid to feed on. Besides the sperm whales, long-finned pilot whales sometimes come in to dive for squid, and occasionally orcas also swim through the area.

Tour operators use directional hydrophones to locate sperm whale clicks. They then head for the sounds and wait for the whales to surface. After about an hour watching the whales, the tour boats typically take a swing inshore to look for Hector's or dusky dolphins. And if they cannot be found, there is always a visit to the seal colonies, on or near the rocks.

Additional places where common, bottlenose, and other dolphin watching occurs, include South Island ports such as Picton and Te Anau. On the North Island, there are tours in the Bay of Islands, accessible through Paihia, and in the Bay of Plenty, accessible through Tauranga and Whakatane. Orcas and long-finned pilot whales are also sometimes seen in these areas.

Experienced whale watchers are always on the lookout for rare beaked whales around New Zealand—a good place to find them. Unfortunately, most beaked whale records are of strandings. New Zealand has some of the highest numbers of cetacean strandings in the world.

SPECIAL FEATURES

Besides sperm whales, New Zealand offers the rare chance to see a small, attractive dolphin with a rounded dorsal fin—the Hector's dolphin. This dolphin is found only in waters around New Zealand. It belongs to the rare genus of *Cephalorhynchus* dolphins with just four members; two of the others are seen off South America, and one off South Africa. Even most cetacean scientists have never seen these dolphins at sea. Perhaps a handful of people worldwide have seen all four. Kaikoura is one place to see Hector's dolphins, but they are even more commonly found farther south, from Akaroa on the Banks Peninsula.

TRAVELER'S NOTES

When to visit *Year-round sperm whales, Hector's, common, and bottlenose dolphins; Oct–May dusky dolphins close to shore (esp. Kaikoura)*
Weather *Cool (summer) to cold (winter) on the sea, Kaikoura, South Island and southern North Island; cool (winter) to warm (summer) on the sea from Bay of Plenty to Bay of Islands, North Island*
Types of tours *Half- and full-day tours; extended multi-day expeditions;* *inflatables, sailboats, motorboats, large whale-watch boats; some whale watching from ferries; helicopter and fixed-wing aircraft*
Tours available *South Island: Kaikoura, Akaroa, Picton, and Te Anau; North Island: Paihia, Tauranga, and Whakatane*
Information *Canterbury Visitor Information Centre, PO Box 2600, Christchurch, New Zealand, Ph. 64 3 379 9629*

Tonga and Other Pacific Islands

Australia and Oceania

The Kingdom of Tonga, the "Friendly Islands," is located in the western South Pacific, 1,400 miles (2,250 km) northeast of New Zealand. Like most South Pacific nations, it is composed mainly of water. Of about 170 islands, only some 36 are inhabited. There are three main island groups: two are coral formations, while the third, Vava'u, is volcanic.

Humpback whales return in large numbers from the Antarctic to the warm seas around Vava'u to mate and calve during the austral winter. Access to the Vava'u group is through the airport north of Neiafu, on 'Uta Vava'u. During the whale-watch season, whales around Vava'u can be seen fighting over potential mates. Even before humpbacks are sighted, a hydrophone is often dropped below the surface of the water to eavesdrop on the singing males.

Whale-watch trips are offered out of the Bounty Bar in Neiafu. Besides humpbacks and occasional sperm whales, there are short-finned pilot whales and spinner and other tropical dolphins that might approach the boat. Whale watching and research are new here, so future trips may reveal more resident or migratory ceatceans.

Several hundred miles northwest of Tonga lies the much more extensive and populous Republic of Fiji, with 844 islands and islets, about 100 of which are inhabited. The larger islands, including

A humpback surfaces in the waters off Tahiti (top). Spinner dolphins (above) are found in many parts of the Pacific Ocean.

the two main ones, are rugged and mountainous. The Great Sea Reef of coral stretches for 300 miles (500 km) along the western fringe of Fiji. Sailboat, kayak, and diving tours are available, offering sightings of tropical dolphins. In 1995, photo-identification and behavioral research began with a small pod of spinner dolphins living near Tavarua and Namotu islands, better known as destinations for serious surfers. These small islands, west of Nadi at the southern edge of the reef that encloses the Mamanucas, are accessible by boat from Nadi, on Viti Levu. Researchers estimate that 60 dolphins live in the area and, according to islanders, dolphins have lived there as long as

anyone can remember. Tourists are invited to spend a day cruising on a boat specially designed for dolphin watching and research.

Cetaceans can also be seen in French Polynesia (Tahiti), New Caledonia, Niue, and Western Samoa. Exciting spinner dolphin tours, begun by an American scientist studying the spinner's behavior, depart from the Richard Gump South Pacific Biological Station on Cook's Bay, on the north side of Moorea, and take visitors outside the lagoon to meet the dolphins. In New Caledonia, dolphin tours are available out of Noumea, and humpback whales are sometimes seen in August and September. There is occasional humpback whale watching from Niue and Western Samoa, but tours are yet to become as popular as in Tonga.

Fiji, "the soft coral capital of the world," has year-round diving only 10–15 minutes by boat from most resorts. Fiji also has live-aboard dive boats, and resorts cater to low- and high-budget divers. Some of the interesting species to be seen include hundreds of hard and soft coral, sea fans, and sea sponges. You might meet the territorial clownfish, living in a symbiotic relationship with the poisonous sea anemone, and surgeonfish, trumpetfish, red lizard fish, batfish, and parrotfish. At the edges of the reefs are barracuda, jackfish, small reef sharks, stingrays, and large parrotfish. Large sharks and the aggressive gray reef shark are usually found away from the coast in deeper water. But remember: coral is alive and fragile—only travel with ecologically sound operators and follow the rules.

SPECIAL FEATURES

No matter which island you visit, don't miss the coral reef diving. The greatest biological diversity in the tropical South Pacific is invariably underwater. If you can't scuba dive, try snorkeling.

TRAVELER'S NOTES

When to visit July–Nov humpback whales in Tonga; year-round, tropical dolphins in all locations, but Apr–Oct best for weather

Weather Hot during dry season Apr–Oct; sun averages 6–8 hours a day, even in wet season

Types of tours Half- and full-day tours, some extended multi-day expeditions; inflatables, sailboats, kayaks, and diving boats

Tours available Tonga: Vava'u; Fiji: Nadi; French Polynesia: Moorea; New Caledonia: Noumea

Information Tourism Council for the South Pacific, 35 Loftus Street, PO Box 13119, Suva, Fiji Islands, Ph. 679 304 t77; Whale Watch Vava'u Ltd., c/- Bounty Bar, Private Bag, Neiafu, Vava'u, Kingdom of Tonga, Ph. 676 70 576

The Vava'u group (above) comprises a part of some 170 islands making up Tonga.

THE POLES

Against stark, frozen vistas, the poles provide a variety of spectacular sights for whale watchers, with an explosion of life throughout the ecosystem.

Before the long history of whaling began, the Arctic and Antarctic regions, in summer, held the greatest density of whales on Earth. Blue, fin, sei, minke, humpback, along with other great whales, gathered in huge numbers around the poles to feed on massive concentrations of krill and other plankton, fish, and squid. During the winter months, most whale species retreated, some migrating as far as the equator to give birth, mate, and raise their calves.

But whaling changed all that. Arctic whaling commenced in 1607 off Svalbard, north of Norway, and spread from there. The bowhead whale was the first Arctic casualty, with many other species following.

Despite its geographic position at the top of the continents of Europe, Asia, and North America, the Arctic has been only a little more accessible for tourism than Antarctica. Traveling around the Arctic is difficult and expensive. The first whale-watch trips, established in the early 1980s, were from North America, to the subarctic of northern Manitoba, Canada, where belugas and polar bears can be seen in and around the Churchill River. Since then, trips to see belugas and narwhals in northern Canada and Greenland have been organized. Since the end of the Cold War, former Russian icebreakers have been made available for extended

Whale watchers in the Antarctic (below) on a large ship.

South America

Falkland Islands

South Georgia Island

South Africa

Tierra del Fuego

Antarctic Peninsula

SOUTHERN OCEAN

Antarctica

SOUTHERN OCEAN

New Zealand

Australia

Scanning the ice floes in Antarctica (left) for orcas. Belugas (below) gather to swim and molt in the freshwater shallows off the coast in the Canadian Arctic.

Siberia

Russia

Novaya Zemlya

Franz Josef Land

Europe

Svalbard

The Arctic

ARCTIC OCEAN

Greenland

Alaska

Baffin Island

Canada

nature cruises to the far north of Canada, Greenland, Norway, and Russia. These boats stop in various ports during their northern journeys, often encountering bowheads, narwhals, and belugas—three cetacean species found only in Arctic regions.

For a long time, the Antarctic, in contrast, seemed to be relatively safe from the threat of whaling. But with the advent of exploding harpoons and long-range factory ships, modern whaling took its toll.

In the first half of the twentieth century, up until the 1960s, hundreds of thousands of whales died in the Antarctic, as one large whale species after another was hunted nearly to extinction. Today, the whales in this region are largely protected from whaling, and some species are slowly rebuilding their numbers. But it will take years before we know if endangered blue whales, for example, will survive, much less return to anything approaching their original numbers.

When the first cruise ship visited the Antarctic Peninsula in 1957, whale watching was unheard of. By the 1980s however, it had become a key feature of organized tours, along with viewing penguins, seals, and giant icebergs. Antarctic cruises offer half a dozen cetacean species seen only in the far south of the Southern Hemisphere, in addition to most of the large baleen whales and the ubiquitous orca.

The Arctic

The Poles

Arctic whales, including tusked narwhals, snow-white belugas, and massive bowhead whales, hold a mystery all their own. For many years, only scientists and adventurers could witness these arctic sea mammals in their icy, year-round habitat. However, now there are organized tours to meet them, mainly during the long summer days of the midnight sun.

True arctic whale watching is accessible through much of the Arctic. Many of these trips are cruises, mainly aboard icebreakers or specially strengthened ships, with itineraries spanning two or more territories, including Greenland, northern Canada, northern Alaska, Svalbard (the far northern island group off northern Norway), Siberia, and the Russian islands of Franz Josef Land and Novaya Zemlya. Such trips are not solely for whale watchers, but they include regular sightings of arctic whales, as well as (depending on location) minke, fin, and humpback whales and orcas. Gray whales can be seen in the eastern Russian and Alaskan Arctic. Closer to land or

ice, you can sometimes see polar bears, walruses, seals, arctic foxes, and many bird species. Helicopters and inflatable Zodiac boats, which are carried on some ships, are also used to find and view whales and other wildlife up close.

Summer is the season various baleen whales come to the Arctic, or subarctic, to feed. At that time, some belugas move into arctic river mouths, while narwhals are seen offshore, segregated into all-male and nursery groups.

Besides cruises, it is possible to fly to remote arctic outposts in northern Canada, Svalbard, and Greenland to take excursions in small boats. One of the better areas is in northwestern Greenland, north of Disko Bay, where whale-watch safaris into Uummannaq Fiord and Baffin Bay depart from Uummannaq. The Uummannaq Tourist Service offers four- to five-hour tours, with hydrophones, and a virtual certainty in summer of seeing and

An impressive show of tusks by male narwhals in the cold waters surrounding the eastern Canadian Arctic (left).

Belugas in the shallows of Somerset Island, in the Canandian Arctic (above). Hudson Bay, near Churchill, Manitoba (right).

hearing whales. Fin whales are the most common, followed by minke whales, narwhals, and belugas, and occasionally sperm whales and orcas. Some of these species can be more easily seen in other parts of the world, but this is one of the few places to find the legendary narwhal, the animal whose tusks were once sold as unicorn horns. Imagine crossed tusks protruding from the water, and the dull thud of tooth against tooth as males joust. (For more information on Greenland, see pp. 236–7.)

Northern Canada also has whale-watch tours from sites around Pond Inlet and Somerset Island to see belugas, narwhals, and bowheads. These are rugged, multi-day tours offered by charter or through expedition tour companies. Some are led by Inuit or local people. There are even whale-watch expeditions by sea kayak.

SPECIAL FEATURES

Perhaps the most accessible sub-arctic location is Churchill, on Hudson Bay, northern Manitoba, Canada. Churchill is accessible by jet or 40-hour passenger train from Winnipeg. Although considerably south of the Arctic Circle, belugas come in close to the mouth of the Churchill River, swimming some miles upriver from June through

August. Churchill is also known as the "polar bear capital of Canada." Polar bears wander into the area from early July. By September, when the belugas leave the river mouth to head back to the open Arctic, the polar bears reach large numbers that peak in late October and early November. August is the best month for those who want the chance to see both belugas and polar bears. All tours should be booked in advance in Churchill.

TRAVELER'S NOTES

When to visit June–Aug for most of the Arctic; Aug best for belugas and polar bears in Churchill

Weather Cold on water throughout the Arctic, snow possible; Churchill, warm temperatures by day, often dipping to freezing at night

Types of tours Half- and full-day tours, some extended multi-day expeditions; inflatables, sailboats, and large whale-watch boats

Tours available Canada: Churchill; Greenland: Uummannaq; Norway: Sveagruva on Svalbard

Information Greenland Tourism a/s, Postbox 1552, DK-3900 Nuuk, Greenland, Ph. 299 22888; Churchill Nature Tours, PO Box 429, Erickson, Manitoba R0J 0P0, Canada, Ph. 1 204 636 2968; Marine Expeditions Inc., 30 Hazelton Avenue, Toronto, Ontario M5R 2E2, Canada, Ph. 1 416 964 9069; Noble Caledonia Ltd., 11 Charles St., Mayfair, London W1X 8LE, England Ph. 44 171 409 0376; Arcturus Expeditions Ltd., PO Box 850, Gartocharn, Alexandria, Dunbartonshire G83 8RL, Scotland, Ph. 44 1389 830 204

Antarctica

The Poles

In many ways the ultimate whale-watch destination, Antarctica is a marine paradise for the nature lover. Most of the cold, deep water that fills the world's ocean basins comes from here. Massive upwellings around the continent create an environment ideal for plankton, which supports the largest and most massive aggregations of krill in the sea.

During the first half of the twentieth century, up until the 1960s, the seas around Antarctica were the scene of large-scale whale slaughter. Even today, several decades after factory whaling has stopped, blue, fin, humpback, and sei whales still exist in far fewer numbers, with the minke whale being the sole exception.

The only way for most people to get to Antarctica and watch whales is aboard a large cruise ship. These trips are not exclusively whale-watch tours but extended cruises that include sightings of birds, seals, icebergs, penguins, and cetaceans, as well as visits to Antarctic research stations. The cruises are offered by various companies based in the United States, Canada, Britain, Germany, Australia, and New Zealand. However, tour agencies worldwide will sell space on the ships. There are a few air tours to Antarctica, but these are not recommended for whale watching.

Besides South American departures, some cruises leave from various ports in Australia,

An orca with calf (above), possibly on the lookout for penguins. Orcas are known to knock penguins off ice floes as they float by.

New Zealand, and South Africa. However, for most visitors, the main part of the journey starts in southern South America, when they board a cruise ship in Ushuaia, Argentina, or nearby southern Chile. The journey from South American ports to Antarctica takes several days, during which the ship passes through three ocean biomes, or ecological zones (see p. 16).

The first zone is along the inshore waters around Tierra del Fuego and the surrounding islands, as the ship winds its way south past Cape Horn. Here you might see Commerson's and Peale's dolphins, and with any luck, some of the rare Burmeister's or spectacled porpoises—all found only, or mainly, in this part of the

A humpback (above) around the Antarctic Peninsula. A meeting between a tour ship and a yacht in Lemaire Channel (left).

to the main continental part of Antarctica, which is 97 percent covered with ice. However, if you can include the main part of the continent, you will see more wildlife and get a glimpse of life nearer the heart of this last unknown continent in an atmosphere that is eerie and majestic.

SPECIAL FEATURES

Some cruises stop in the Falkland Islands either before heading south for Antarctica, or on the return journey. Watch for the orcas that often feed on gentoo penguins around Sea Lion Island. Other cruises visit the subantarctic island of South Georgia, site of a large former whaling station, as well as home to fur seals, elephant seals, breeding king and macaroni penguins, and albatrosses.

world—as well as two Southern Hemisphere residents, the dusky dolphin and the southern bottlenose whale. As you motor across the Drake Passage and south to Antarctica, reaching the high seas off the continental shelf, hourglass dolphins sometimes accompany the ship, and you might see long-finned pilot whales and southern right whale dolphins. Keep a lookout for fin, sei, and Cuvier's beaked whales, too. Finally, as the Antarctic Peninsula nears, minke and humpback whales, which migrate closer to the ice, are more commonly seen feeding in the rich patches of krill and fish. Moving through the last two zones are opportunistic orcas, feeding on fish and seals and even knocking penguins from ice floes at times. This is life on a larger canvas than almost anywhere else in the world.

Most trips visit the Antarctic Peninsula; only the more expensive, longer cruises venture

TRAVELER'S NOTES

When to visit *Late Nov–Mar: baleen whales—humpback, minke, fin—feed around Antarctica, along with orcas and hourglass dolphins. The window for sailing to Antarctica is only 3–4 months*

Weather *Cold, take winter gear even in Antarctic summer; cruise ships offer shelter from rain and cold, but dress warmly to see whales close up from deck or inflatable boats launched from the ship*

Types of tours *Extended multi-day expeditions*

Tours available *Through Abercrombie & Kent (USA), Adventure Associates*

(Australia), Aurora Expeditions (Australia), Hanseatic Tours GMBH (Germany), Marine Expeditions Inc. (Canada), Mountain Travel-Sobek (USA), Natural Habitat Adventures (USA), Ocean Adventures (UK), Orient Lines (USA), Quark Expeditions (USA), Society Expeditions (USA), Southern Heritage Expeditions (New Zealand), Special Expeditions (USA), Travel Dynamics (USA), Wild-Oceans/WildWings (UK), and Zegrahm Expeditions (USA). Some cruise ships carry inflatables for close viewing

Information *Contact your travel agent*

Nature and books belong to the eyes that see them.

Essays: Second Series

RALPH WALDO EMERSON (1803–82), American philosopher, essayist, and poet

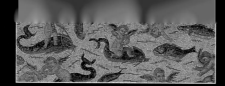

RESOURCES
DIRECTORY

FURTHER INFORMATION

Books

Among Whales, by Roger Payne (Scribner, 1995).

Arctic Whales, by Stefani Payne (Greystoke Books, 1995).

The Book of Dolphins, by Mark Carwardine (Dragon's World, 1996). A valuable reference source on dolphins, with the latest scientific research, up-to-date information, anecdotes, and first-hand encounters.

Cetacean Behavior: Mechanisms and Functions, edited by L. M. Herman (J. Wiley, 1980).

The Conservation of Whales and Dolphins: Science and Practice, edited by Mark P. Simmonds and Judith D. Hutchinson (J. Wiley, 1996).

Dolphin Days, by Kenneth S. Norris (Norton, 1991).

Dolphin Societies, by Karen Pryor and Kenneth S. Norris (University of California Press, 1991).

Dolphins, by Chris Cotton (Boxtree, 1995).

Dolphins, Porpoises and Whales of the World, The IUCN Red Data Book, by Margaret Klinowska (IUCN, 1991). Detailed conservation reports and general species information on all cetaceans.

The Ecology of Whales and Dolphins, by D. E. Gaskin (Heinemann, 1982).

A Field Guide to Whales, Porpoises and Seals from Cape Cod to Newfoundland, 4th Edition (revised), by Steven K. Katona, Valerie Rough, and David T. Richardson, (Smithsonian Institution, 1993).

The Greenpeace Book of Dolphins, edited by John May (Sterling, 1990).

Guardians of the Whales: the Quest to Study Whales in the Wild, by Bruce Obee and Graeme Ellis (Whitecap Books, 1992).

A Guide to the Photographic Identification of Individual Whales, by Jon Lien and Steven Katona (American Cetacean Society, 1990). A species-by-species guide to identifying individual cetaceans based on their natural and acquired markings.

Handbook of Marine Mammals: Vol. III, 1985; Vol. IV, 1989; Vol. V, edited by Sam H. Ridgway and Sir Richard Harrison (Academic Press, 1994). Detailed accounts of cetacean species, along with literature surveys.

A History of World Whaling, by Daniel Francis (Viking, 1990).

Marine Mammal Sensory Systems, edited by J. A. Thomas, R. A. Kastelein, and A. Y. Supin (Plenum, 1992).

Marine Mammals of the World, FAO Species Identification Guide, by Thomas A. Jefferson, Stephen Leatherwood, and Marc A. Webber (UNEP, FAO, 1993). A good overall guide to marine mammals, including cetaceans, with identifying external views as well as skull and diagnostic jaw illustrations.

The Natural History of Whales and Dolphins, by P. G. H. Evans (Christopher Helm, 1987).

On the Trail of the Whale, by Mark Carwardine (Thunder Bay Publishing Co., 1994). Covers an adventurous year crisscrossing the globe in search of whales, from the North Atlantic to the South Pacific.

Orca: The Whale Called Killer, by Erich Hoyt (Firefly, 1990). A thorough book outlining the social behavior and biology of orcas; based on a narrative covering seven summer expeditions spent living with orcas around Vancouver Island, Canada.

Sealife: A Complete Guide to the Marine Environment, edited by Geoffrey Waller (Pica Press, 1996).

Seasons of the Whale: Riding the Currents of the North Atlantic, by Erich Hoyt (WDCS/Humane Society, 1998). An illustrated account of an eventful year in the lives of several humpback, right, and blue whales known to both whale watchers and researchers.

The Sierra Club Handbook of Whales and Dolphins, by Stephen Leatherwood and Randall R. Reeves (Sierra Club Books, 1983).

Voyage to the Whales, by Hal Whitehead (Robert Hale, 1989).

The Whale Watcher's Guide, by Patricia Corrigan (North-Word Press, 1994).

The Whale Watcher's Handbook, by Erich Hoyt (Doubleday, 1984). A definitive guide to whale and dolphin watching around the world.

Whales, by W. Nigel Bonner (Blandford, 1980).

Whales, by Jacques-Yves Cousteau and Yves Paccalet (Harry N. Abrams, 1986).

Whales and Dolphins, by Anthony R. Martin (Salamander Books, 1990).

Whales, Dolphins and Porpoises, by James D. Darling, C. Nicklin, K. S. Norris, H. Whitehead, and B. Wursig (National Geographic, 1995).

Whales, Dolphins and Porpoises, edited by Michael M. Bryden and Sir Richard J. Harrison (Merehurst Press, 1988).

Whales, Dolphins and Porpoises, by Mark Carwardine (Dorling Kindersley, 1995). A definitive field guide to all the world's cetaceans; packed with comprehensive information on identification and behavior.

Whales, Whaling and Whale Research: a selected bibliography, by L. R. Magnolia (The Whaling Museum, Cold Spring Harbor, Long Island, New York, 1977).

With the Whales, by James Darling and Flip Nicklin (NorthWord Press, 1990).

Magazines & Journals

BBC Wildlife, BBC Magazines, United Kingdom
Defenders, Defenders of Wildlife, USA
Equinox, Canada
Marine Mammal Science, Society for Marine Mammalogy, USA
National Geographic Magazine, USA
Ocean Realm, Friends of the Sea, USA
Sonar, Whale and Dolphin Conservation Society, United Kingdom
The Whalewatcher, American Cetacean Society, USA

Websites

American Cetacean Society.
http://www.acsonline.org/
The oldest whale conservation group in the world.
A non-profit, volunteer organization devoted to education in the field of cetacean research.
Dept. of Vertebrate Zoology, National Museum of Natural History, Smithsonian Institution.
http://nmnhwww.si.edu/departments/vert.html
Covers the issues surrounding the study of marine mammals as a part of the Marine Mammal Program.
Institute of Cetacean Research.
http://www.whalesci.org/
Ocean Link.
http://oceanlink.island.net/olink.html
Project Aware (Aquatic World Awareness, Responsibility, and Education).
http://www/padi.com/
Details of environmental and educational programs to help preserve and protect the underwater environment.
Society for Marine Mammalogy.
http://pegasus.cc.ucf.edu/~smm/
Whale and Dolphin Conservation Society.
http://www.wdcs.org/wdcs/index.htm
The leading voice for whales and dolphins worldwide; initiates and supports cetacean research and conservation.

WhaleNet.
http://whale.wheelock.edu/
An educational link focusing on whales and marine research.
WWF Global Network.
http://www.panda.org/
A comprehensive information source connected to the world's largest and most experienced independent conservation organization.

Video Cassettes and CD-ROMs

The Free Willy Story, Keiko's Journey Home, (Discovery Communications, Inc., 1996). Discovery Channel Production, produced in association with ABC/Kane Productions. Genesis award winner, narrated by Rene Russo.
In the Company of Whales, (Discovery Communications, Inc., 1995). Discovery Channel CD-ROM for PC, Dr. Roger Payne, narrated by Patrick Stewart. Includes 45 minutes of video, more than 200 compelling photographs, hypertext glossary, and much more.
In the Company of Whales: Gentle Giants of the Watery Realm, (Discovery Communications, Inc., 1992). Discovery Channel video produced in association with Channel Four Television, Television New Zealand. The scientific advisor and host of this video is Dr. Roger Payne, a world renowned cetacean specialist, with narration by Jessica Tandy.
Marine Mammals of the World, CD-ROM, T. A. Jefferson, S. Leatherwood, and M. A. Webber (Springer-Verlag, 1996).
Pacific Blue: Musical Soundscapes (I and II), (Holborne Distribution Co. Ltd.).
Songs of the Whales and Dolphins, CD-ROM, (Marine Mammal Fund/The Nature Company, 1993).
Whale Symphony, CD-ROM, (Ocean Studio).
Whales and Dolphins of the World, CD-ROM, (Webster Publishing, 1996).
World of Whales, CD-ROM, (Teramedia, 1996).

Museums
USA

California Academy of Sciences, San Francisco, California
Cold Spring Harbor Whaling Museum, Cold Spring Harbor, Long Island, New York
Museum of Comparative Zoology, Cambridge, Massachusetts
National Museum of Natural History, Smithsonian Institution, Washington, DC
Natural History Museum of Los Angeles, Los Angeles, California
Scripps Institution of Oceanography, University of California, San Diego, California

AUSTRALIA

Australian Museum, Sydney, New South Wales
Queensland Museum, Brisbane, Queensland
South Australian Museum, Adelaide, South Australia
Western Australian Museum, Perth, Western Australia

CANADA

BC Provincial Museum, Victoria, British Columbia
National Museum, Ottawa, Ontario
Newfoundland Museum, St John's, Newfoundland
Nova Scotia Museum, Halifax, Nova Scotia

JAPAN

National Science Museum Tokyo, Tokyo

THE NETHERLANDS

National Museum of Natural History, Leiden

SOUTH AFRICA

Port Elizabeth Museum, Humewood
South African Museum, Cape Town

SWITZERLAND

Zoologisches Museum der Universitat Zurich, Zurich

UNITED KINGDOM

Natural History Museum, London, England
Royal Museum of Scotland, Edinburgh, Scotland

ORGANIZATIONS

American Cetacean Society,
PO Box 1391,
San Pedro, CA 90731,
USA;
tel. (310) 548 6279
fax (310) 548 6950
e-mail: acs@pobox.com

Audubon Ecology Camps and
Workshops,
National Audubon Society,
613 Riversville Rd,
Greenwich, CT 06831,
USA;
tel. (203) 869 2071
e-mail: webmaster@list.
audubon.org

Australian Antarctic Division,
Channel Highway,
Kingston, Tas. 7050,
Australia;
tel. (61 3) 6232 3209
fax (61 3) 6232 3288

Australian Whale
Conservation Society,
PO Box 12046,
Elizabeth St, Brisbane,
Qld. 4002,
Australia;
tel. (61 7) 3398 2928
fax (61 7) 3398 2215

Center for Coastal Studies,
PO Box 1036,
Provincetown, MA 02657,
USA;
tel. (508) 487 3622
fax (508) 487 4495
e-mail: ccswhale@wn.net

Center for Whale Research,
1359 Smugglers Cove,
PO Box 1577,
Friday Harbor,
WA 98250-0157,USA;
tel. (360) 378 5835

Cetacean Research Unit,
Gloucester Fishermans
Museum,
PO Box 159,
Gloucester, MA 01930,
USA;
tel. (508) 281 6351
fax (508) 281 5666
e-mail: info@cetacean.org

Cetacean Society
International,
PO Box 953,
Georgetown, CT 08329,
USA;
tel./fax (203) 544 8617
e-mail: 73577.310@
compuserve.com

The Cousteau Society,
870 Greenbrier Circle,
Suite 402, Chesapeake,
VA 23320-2641,
USA;
tel. (757) 523 9335
fax (757) 523 2747

The Dolphin Project,
PO Box 8436,
Jupiter, FL 33468,
USA;
e-mail: wilddolphin@
econet.igc.org

The Dolphin Research
Project, Inc.,
PO Box 1245,
Frankston, Vic. 3199,
Australia;
tel. (61 3) 9783 7466
fax (61 3) 9783 7048
e-mail: dolresin@iaccess.
com.au

Dolphin Society,
PO Box 2052,
Clovelly,
NSW 2031,
Australia;
tel. (61 2) 9665
0712
fax (61 2) 9664 2018
e-mail: 100352.2334@
compuserve.com

Earth Island Institute,
300 Broadway,
Suite 28,
San Francisco, CA 94123,
USA;
tel. (415) 788 3666
fax (415) 788 7324
e-mail: earthisland@
earthisland.org

Earthtrust,
25 Kaneohe Bay Drive,
Suite 205,
Kailua, HI 96734,
USA;
tel. (808) 254 2866
fax (808) 254 6409
e-mail: earthtrust@aloha. net

Earthwatch,
680 Mt. Auburn St,
PO Box 403,
Watertown, MA 02272,
USA;
tel. (617) 926 8200
fax (617) 926 8532
e-mail: info@earthwatch.org

European Cetacean Society,
Secretary: Beatrice Jann
Museo Cantonale di Storia
Naturale,
Vialle Cattaneo 4,
CH-6900, Lugano,
Switzerland;
tel./fax (41 91) 966 0953
e-mail: Jann@dial.eunet.ch

Friends of the Earth,
Global Building,
Suite 300,
1025 Vermont Ave NW,
Washington, DC 2005,
USA;
tel. (202) 783 7400

Fundacion Cethus,
PO Box 4490-(1001),
Buenos Aires,
Argentina;
fax (54 1) 823 9739

Great Whales Foundation,
Box 6847,
Malibu, CA 90264,
USA;
tel. (415) 458 3262
fax (310) 317 1414
e-mail: francis@elfi.com

Greenpeace,
1436 U St NW,
Washington, DC 20009,
USA;
tel. (202) 462 1177
fax (202) 462 4507
e-mail: info@wdc.
greenpeace.org

The Humane Society of the
United States (HSUS),
2100 L St NW,
Washington, DC 20037,
USA;
tel. (202) 452 1100

International Dolphin Watch,
Parklands,
North Feriby,
Humberside HU14 3ET,
England;
tel. (44 1482) 643 403
fax (44 1482) 634 914

International Fund for
Animal Welfare,
PO Box 193,
411 Main St,
Yarmouth Port,
MA 02675,
USA;
tel. (508) 362 2649
e-mail: ifaw@easynet.
co.uk

International Whaling
Commission,
The Red House,
Histon,
Cambridge CB4 4NP,
England;
tel. (44 1223) 233 971

International Wildlife
Coalition,
634 North Falmouth
Highway,
North Falmouth,
MA 02556-0388, USA;
tel. (508) 548 8328
fax (508) 548 8542
e-mail: iwcadopt@ccsnet.
com

International Wildlife
Coalition/Brazil,
PO Box 5087,
Florianopolis,
SC 88040-970, Brazil;
tel. (55 48) 234 0021
fax (55 48) 234 1580

Mingan Island Cetacean Study,
Contact (during winter):
Richard Sears
285 Green St,
St Lambert, QC J4P 1T3,
Canada;
tel. (514) 948 3669
fax (514) 948 1131
e-mail: micshipolari@
videotron.ca
Contact (during summer):
Richard Sears
124 Bord de la Mer,
Longue-Pointe-de-Mingan,
QC G0G 1V0,
Canada;
tel. (418) 949 2845

National Marine Fisheries
Service,
National Marine Mammal
Laboratory,
7600 Sand Point Way NE,
Seattle, WA 98115, USA

Northwest Fisheries Science
Center,
2725 Montlake Blvd E,
Seattle, WA 98112,
USA;
tel. (206) 860 3200

Oceanic Society Expeditions,
Fort Mason Center,
Building E, Room 230,
San Francisco,
CA 94123-1394,
USA;
tel. (415) 474 3385
fax (415) 441 1106

The Pacific Cetacean Group,
UC MBEST Center,
3239 Imjin Rd,
Marina, CA 93933,
USA;
tel. (408) 582 1030
e-mail: P007GC@aol.com.

Pacific Whale Foundation,
Kealia Beach Plaza,
Suite 25,
101 North Kihei Rd,
Kihei, Maui, HI 96753,
USA;
tel. (808) 879 8811

Society for Marine
Mammalogy,
PO Box 368,
Lawrence, KS 66044, USA

Southwest Fisheries Science
Center,
PO Box 271,
8604 La Jolla Shores Drive,
La Jolla, CA 92038-0271,
USA;
tel. (619) 546 7000
fax (619) 546 7003
e-mail: Maryann.Rodriguez
@noaa.gov

Tethys Research Institute,
Dr. Giuseppe Nortarbartolo
di Sciara,

Acquario Civico - viale G.B.,
Gadio 2, I-20121 Milano,
Italy;
tel. (39 2) 7200 1947
fax (39 2) 7200 1946
e-mail: gnstri@
imiucca.csi.unimi.it

Whale and Dolphin
Conservation Society,
Alexander House,
James St W,
Bath BA1 2BT,
England;
tel. (44 1225) 334 511
fax (44 1225) 480 097
e-mail: 100417.1464@
compuserve.com

The Whale Center,
8480 Andenes,
Norway;
fax (47 761) 1 5610
e-mail: ancru@nord.
eunet.no

The Whale Museum,
62 First St N,
PO Box 945,
Friday Harbor,
WA 98250,
USA;
tel. (360) 378 4710
fax (360) 378 5790
e-mail: whale@rockisland.
com

WhaleNet,
200 The Riverway,
Boston, MA 02215,
USA;
tel. (617) 734 5200
fax (617) 566 7369
e-mail: williams@whale.
wheelock.edu

The World Conservation
Union (IUCN),
Suite 502,
1400 16th St NW,
Washington, DC 20036,
USA;
tel. (202) 797 5454
fax (202) 797 5461

World Wildlife Fund,
1250 24th St NW,
Suite 500,
Washington,
DC 20037-1175,
USA;
tel. (202) 293 4800
e-mail: wwfus@
worldwildlife.org

World Wide Fund for Nature
International,
Avenue du Mont-Blanc,
CH-1196, Gland,
Switzerland;
tel. (41 22) 364 9111

INDEX *and* GLOSSARY

I n this combined index and glossary, bold page numbers indicate the main reference, and italics indicate illustrations and photographs.

CONTRIBUTORS

Mark Carwardine graduated in zoology from London University and worked as Conservation Officer with the World Wide Fund for Nature in Britain. He has also worked as a science writer with the United Nations Environment Programme (UNEP) in Nairobi, Kenya, and as a consultant to the World Conservation Union in Switzerland. Since 1986, Mark has been a freelance consultant, writer, lecturer, and broadcaster. He has written extensively for both children and adults, and his many books include *Last Chance to See* (with Douglas Adams, *On the Trail of the Whale*, and *The Guinness Book of Animal Records*. He resides in the United Kingdom.

Ewan Fordyce completed his doctorate in zoology at the University of Canterbury. He then carried out research at the Smithsonian Institution and at Monash University in Melbourne. More recently, he has become involved in the study of fossil cetacea from Antarctica, Australia, and New Zealand, and the recovery of scientific material from strandings. Ewan is also keenly interested in the taxonomy, anatomy, and the evolution of fossil and living cetacea. He is Associate Professor, Department of Geology at the University of Otago, New Zealand and a research associate of the Smithsonian Institution.

Peter Gill has been involved in the study of whales since 1983. He has an honors degree in zoology from Flinders University, Adelaide. Peter has undertaken extensive research in Antarctica, Australia, and New Caledonia, observing whales (especially humpbacks and southern rights) at close range. Peter spent two years working for Greenpeace as a marine mammals researcher. He presently works in the Blue Mountains, Australia, as a freelance marine mammal researcher, photographer, lecturer, and writer, and is currently involved in a major program of Antarctic cetacean research.

Erich Hoyt has been going to sea since 1973 when, as a documentary filmmaker, he met orcas off the west coast of Canada. His books include *Orca: The Whale Called Killer*, and *The Earth Dwellers*. His articles on wildlife, conservation, and science have appeared in more than 150 magazines and newspapers, including *National Geographic* and *The New York Times*. Erich has helped design exhibits at science museums in the United States. Since 1990, he has been a consultant for the Whale and Dolphin Conservation Society, writing and speaking on captive dolphins, whale watching, and marine-protected areas. He resides in Scotland.

CAPTIONS

Page 1: Beluga whale.
Page 2: Bottlenose dolphins, Bahamas.
Page 3: Humpback whale fluking, Hawaii.
Pages 4–5: Orca spyhopping in pack ice, McMurdo Sound, Victoria Land, Antarctica.
Pages 6–7: Humpback whales breaching.
Pages 8–9: Atlantic spotted dolphins.
Pages 10–11: Pod of common dolphins, Sea of Cortez, Mexico.
Pages 12–13: Beluga whale.
Pages 32–3: Female humpback whale and her week-old calf swimming in the South Pacific Ocean near Tonga.

Pages 58–9: Humpback whales feeding on herring.
Pages 80–1: Breaching orca off Vancouver Island, Canada.
Pages 102–3: A minke whale swims upside down past a tourist boat, Husavik, Iceland.
Pages 122–3: A gray whale passes through kelp, Channel Islands, California.
Pages 142–3: Atlantic spotted dolphins swimming in the Bahamas.
Pages 194–5: Northern right whale displaying its flukes, Newfoundland.
Pages 272–3: A Tunisian mosaic depicting cherubs playing with dolphins.

ACKNOWLEDGEMENTS

The publishers wish to thank the following people for their assistance in the production of this book:
Apple Computer Australia Pty Ltd, Garry Cousins, Margaret McPhee, Paddy Pallin Ltd.

PICTURE AND ILLUSTRATION CREDITS

t = top; b = bottom; l = left; r = right; c = center
Ardea = Ardea London Ltd; **Auscape** = Auscape International; **BCA** = B & C Alexander; **Bob Cranston** = Bob Cranston, San Diego; **Hedgehog** = Hedgehog House New Zealand; **IPL** = International Photo Library; **IV** = Innerspace Visions; **MC** = Mark Carwardine; **Minden** = Minden Pictures; **MMI** = Marine Mammal Images; **OEI** = Ocean Earth Images; **OSF** = Oxford Scientific Films; **PEP** = Planet Earth Pictures; **Stock** = Stock Photos P/L; **TGC** = The Granger Collection, New York; **TIB** = The Image Bank; **TPL** = The Photo Library, Sydney; **TSA** = Tom Stack & Associates; **WO** Weldon Owen Pty Ltd
Picture Credits
1c Kevin Schafer/Hedgehog 2c Doug Perrine/Auscape 3c Doug Perrine/PEP **4–5** Kerry Lorimer/Hedgehog **6–7** Sanford/Agliolo/Stock **8–9** James D. Watt/MMI **10–11** Bob Cranston **12–13** Kevin Schafer/Hedgehog 14t MC; bl James D. Watt/PEP; br Richard Coomber/PEP; bt David B. Fleetham/OSF 15t Doug Perrine/Auscape 16c Colin Monteath/Auscape **16–17**c Becca Saunders/Auscape; 17r François Gohier/Ardea; b Kenneth C. Balcomb III/EarthViews 18c James D. Watt/PEP; bl Ian Beames/Ardea 19b Andrea Florence/Ardea 20c e.t. archive 21t François Gohier/Ardea; c Michael S. Nolan/IV; b Bill Wood/PEP 22b TGC 23c TGC; b Image Select 24t Mary Evans Picture Library; b Museo Nazionale Napoli/Scala 25t Lauros-Giraudon; c François Gohier/Ardea 26t Pieter Folkens/PEP; c Brian Parker/TSA 27t Donald Tipton/Underwater Images; c Brandon D. Cole/IV; **28**t Graham Robertson/Auscape; c Doug Perrine/IV

29bl MC; br MC 30t European Space Agency/SPL/TPL; c Flip Nicklin/Minden 31t Flip Nicklin/Minden; c Michael Freeman/Phototake/Stock **32–33** Jean-Marc La Roque/Ardea 34t Roger-Viollet; b TGC 35t TGC; c Old Dartmouth Historical Society - New Bedford Whaling Museum 36t BCA; bl Musee de Carnavalet, Paris/e.t. archive; br Victoria & Albert Museum/e.t. archive 37t Carl Bento/Nature Focus/Australian Museum; c Bargello Florence/e.t. archive 38t Ann Ronan at Image Select; b Image Library, State Library of NSW 39t Mary Evans Picture Library; c MMI; bl Mary Evans Picture Library; br Ian Cummings/BCA 40b Bob Cranston 41t Brian Sytnyk/Stock; c Kelvin Aitken/OEI; b Bob Cranston 42t Gleizes/Greenpeace; b Wendy Else/BCA 43t Harald Sund/TIB; c McTaggart/Greenpeace 44t BCA; b Edward S. Curtis/National Geographic Image Collection 45b Galen Rowell/Mountain Light Photography 46t Scott Benson/MMI; b Adam Woolfitt/Robert Harding Picture Library 47t BCA; c Robert L. Pitman/IV 48c Tony Arruza/TPL; b MC 49t Charlie Dass/TPL; b MC 50c Wei Dong Cheng/TIB; b Flip Nicklin/Minden 51t BCA 52c Simon Fraser/SPL/TPL 53tl Wyb Hoek/MMI; tr T. Kitchin/TSA; b Tony Bomford/OSF 54t MC; b P. & G. Bowater/TIB 55t Duncan Maxwell/Robert Harding Picture Library; c P. & G. Bowater/TIB 56b MC 57tl Beltra/Greenpeace; tr Geier/Greenpeace; b Steve Dawson/Hedgehog **58–59** Michio Hoshino/Minden 60tl R. Ewan Fordyce; tr R. Ewan Fordyce 61l R. Ewan Fordyce; c R. Ewan Fordyce 62t Marty Snyderman/IV 63t Michael S. Nolan/IV; c D. Parer & E. Parer-Cook/Auscape; b Richard Herrmann/OSF 64t

Tony Martin/OSF; b Peter Gill 65t Doug Perrine/IV; b Tony Martin/OSF 66t Jeff Foott/TSA 67t Brandon D. Cole/IV; c Doug Perrine/IV 68b Flip Nicklin/Minden 69t R. Ewan Fordyce; b Doug Perrine/IV 71tl Mike Osmond/Pacific Whale Foundation/Auscape; tr Phillip Colla/IV; b Michael S. Nolan/IV 72t Marilyn Kazmers/IV; b François Gohier/Ardea 73t Doug Perrine/IV; c Brian Parker/TSA 74b Doug Perrine/IV 75t Ingrid Visser/EarthViews; b Kelvin Aitken/OEI 76t Bob Cranston 77t Peter Gill; b Doug Perrine/IV 79t Jean-Paul Ferrero/Ardea; c Howard Hall/PEP; b François Gohier/Auscape 80–81 Bob Cranston 82t François Gohier/Ardea; c Doug Perrine/IV; b Pieter Folkens/PEP 83t Andrea Florence/Ardea 84t Doug Perrine/Auscape 84–85t Flip Nicklin/Minden; 85b Doug Perrine/IV 86t Rod Scott; c IFAW/IV; b Doug Perrine/IV 87t Doug Perrine/PEP 88t D. Parer & E. Parer-Cook/Auscape; b Doug Perrine/IV 89t François Gohier/Ardea 90t Doc White/PEP; c Pieter Folkens/PEP 91t James D. Watt/PEP; c Clive Bromhall/OSF; b Doug Perrine/Auscape 92c Daniel J. Cox/Liaison International/Wildlight Photo Agency; b Doug Perrine/Auscape 93t Flip Nicklin/Minden; b Alain Ernoult/TIB 94c Doug Perrine/IV; b Howard Hall/OSF 94–95t Doug Perrine/PEP 95c Flip Nicklin/Minden 96l Richard Smyth/Auscape; b Jen & Des Bartlett/OSF 97t Peter Gill; c Doug Perrine/IV 98c Jean-Paul Ferrero/Ardea 98–99t John & Val Butler/Lochman Transparencies; 99c The New Zealand Herald; b Jean-Paul Ferrero/Auscape 100t Tui De Roy/OSF; b Milt Putnam/Stock 101t Doug Perrine/PEP; c Jiri Lochman/Lochman Transparencies 102–103 BCA 104t Diana McIntyre/MMI; b Howard Hall/OSF 105t Colin Monteath/Auscape; b Jiri Lochman/Lochman Transparencies 106tl WO; tr WO; b Paddy Pallin 107t MC; b WO 108t Stuart Bowey/Ad Libitum/WO; b François Gohier/Ardea 109cl WO; cr Michael S. Nolan/IV; b Apple Computer Australia Pty Ltd 110t Michael S. Nolan/MMI; bl Oliver Strewe/WO; br David M. Hamilton/TIB 111t Mark Conlin/PEP; b John Borthwick 112c Alisa Schulman/MMI; 112–113t Steve R. Burnell; b John Eastcott/PEP 113tr Peter Gill 114t MC; b Colin Monteath/Auscape 115t Thomas Kitchin/TSA; b Flip Nicklin/Minden 116t François Gohier/Ardea; c Dover Publishers/Animals; b Marilyn Kazmers/IV 117t Marty Snyderman/PEP; b Flip Nicklin/Minden 118t Kurt Amsler/PEP; c Ben Osborne/OSF 119t Michael S. Nolan/IV; r Mark Carwardine/IV 120bl Carlos Angel/Gamma/Picture Media; br Silvan Wick/MMI 121tl Frank Nowikowski/South American Pictures; tr Jeff Foott/TSA 122–123 Bob Cranston 124t MC; c MC 125c Kevin Deacon/OEI; b Neville Dawson/TPL 126t Chip Matheson/MMI 127t Hiroya Minakunchi/IV; b Barbara Todd/Hedgehog 128tl WO; tr WO; b MC 129tl Peter Gill; tr(inset) Michael Kozicki/MMI; c Doug Perrine/IV 130t Doug Perrine/PEP; c Godfrey Merlen/OSF; b MC 133t MC; c Phillip Colla/IV(#1188); b David B. Fleetham/OSF 134tl D. Parer & E. Parer-Cook/Auscape; tr Thomas Kitchin/TSA; b Mary Evans Picture Library 136c Doug Allan/OSF 137t Howard K. Suzuki/IV; c Robin W. Baird/MMI; b Doug Perrine/PEP 138t Diana McIntyre/MMI; c François Gohier/Auscape; b Flip Nicklin/Minden 139t Jim Nahmens/MMI 140bl Doug Perrine/PEP; br Thomas Jefferson/MMI 141t MC; c Doug Perrine/PEP 142–143 Doug Perrine/PEP 144–145 repeat of pages 176/177 and 188/189 146t Bob Cranston; b Katy Penland/MMI 147t François Gohier/Ardea; b Lura Meyer/MMI 148t Bob Cranston; b James D. Watt/IV 149t Greg Silber/EarthViews 150t Kevin Deacon/OEI; b Ben Osborne/OSF 151t Tui De Roy/Hedgehog 152t Doug Perrine/Auscape 153t Tui De Roy/Hedgehog; c MC 154t Doc White/PEP; c IPL 155t Bob Cranston; b Mike Bacon/TSA 156t François Gohier/Ardea; b Doug Perrine/Auscape 157t Robert L. Pitman/EarthViews 158t Whale Watch Azores/IV 159t Hal Whitehead/MMI; b Godfrey Merlen/OSF 160t Scott Benson/MMI 161t James D. Watt/IV 164t MC 165t Gregory Ochocki/PEP; b Flip Nicklin/Minden 166t Andrea Florence/Ardea; b Wyb Hoek/MMI 167t Thomas Jefferson/MMI 168t Ed Robinson/TSA; b Doug Perrine/IV 169t Doug Perrine/Auscape; c Doug Perrine/Auscape 170t Hal Whitehead/MMI; b Richard Sears/EarthViews 171t Phillip Colla/IV(#44); b Marilyn Kazmers/IV 172t Steve Dawson/Hedgehog; b Hiroya Minakuchi/IV 173t MC; b Paul Ensor/Hedgehog 174t Colin Monteath/Hedgehog; b Colin Monteath/Auscape 175t Robert L. Pitman/IV; c Steve Dawson/Hedgehog 176t Pete Oxford/PEP; b Robert L. Pitman/IV 177t Doug Perrine/IV 178t Robert L. Pitman; b James D. Watt/PEP 179t Robert L. Pitman/IV; b Robert L. Pitman/EarthViews 180t Robert L. Pitman/IV; b James D. Watt/IV 181t Doug Perrine/PEP; c Flip Nicklin/Minden 182t Doug Perrine/Auscape; b Doug Perrine/IV 183t François Gohier/Ardea; c Robert L. Pitman/IV 184c Stephen Leatherwood/EarthViews 185t Dave B. Fleetham/TSA; b Doug Perrine/IV 186t Bob Cranston;

b Michael S. Nolan/EarthViews 187t Kim Westerskov/OSF; c François Gohier/Ardea 188t Kelvin Aitken/OEI; b Doug Perrine/IV 189t Dennis Buurman/Hedgehog 190t Flip Nicklin/Minden; b Doug Allan/OSF 191t Jeff Foott/TSA; c Gary Milburn/TSA 192t Sharon Nogg/MMI; b Robin W. Baird/MMI 193t Kenneth C. Balcomb III/EarthViews; b C. Faesi/MMI 194–195 François Gohier/Auscape 196–197 repeat of pages 208/209 198t Mark Conlin/IV; b Doug Perrine/IV(Hawaii Whale Research Foundation NMES permit #633) 199b Eric Sander/Gamma/Picture Media 200c Brandon D. Cole/IV; b Richard Sears/EarthViews 200–201t Duncan Murrell/OSF 201c Nick Nicholson/TIB 202c Marilyn Kazmers/IV 202–203t Michael S. Nolan/IV 203c François Gohier/Ardea 204c Tom Campbell/IV 204–205t Bob Cranston 205c ACS-G Bakker/MMI 206c Tui De Roy/Hedgehog 206–207t Marilyn Kazmers/IV 207c MC 208cl Howard K. Suzuki/IV; cr Michael S. Nolan/IV 208–209t James D. Watt/PEP 209c Kjell Sandved/OSF 210t Mel Digiacomo/TIB; b Doug Perrine/IV 211c Doug Perrine/IV 212c David B. Fleetham/OSF 212–213t Monica Borobia/MMI 213c Ken Lucas/PEP 214c BCA 214–215t MC; b Godfrey Merlen/OSF 215c Joseph Devenney/TIB 216c MC 216–217t Steve Bunnell/TIB 217c MC; b MC 218c Doug Perrine/PEP 218–219t Marty Snyderman/PEP 219c Doug Perrine/PEP 220c Doug Perrine/Auscape 220–221c MC 221c Doug Perrine/PEP; b MC 222t James D. Watt/EarthViews; b Skeet McAuley/TIB 223c Frank Nowikowski/South American Pictures 224c Carlos Angel/Gamma/Picture Media 224–225c Carlos Angel/Gamma/Picture Media 225c Luis Castaneda/TIB 226c Thomas Henningsen/MMI 226–227t Andrea Florence/Ardea 227c Ann Ripp/TIB 228c Doug Perrine/PEP 228–229c MC 229c Doug Perrine/IV 230t Clive Collins; b Doug Perrine/IV 231t Adam Woolfitt/Robert Harding Picture Library; c MC 232c MC 232–233t Michio Hoshino/Minden 233c BCA 234c Tui De Roy/Auscape 234–235t D. Parer & E. Parer-Cook/Auscape 235c MC 236c Tui De Roy/OSF 236–237t François Gohier/Auscape 237c David Lomax/Robert Harding Picture Library; b Charles Bishop/PEP 238b Dominic Harcourt-Webster/Robert Harding Picture Library 238–239t Andrea Pistolesi/TIB 239c Doug Perrine/IV 240c François Gohier/Ardea 240–241t MC 241c Gianalberto Cigolini/TIB 242c Doug Perrine/Auscape 242–243t MC 243c MC; b Michael Pasdzior/TIB 244c MC 244–245c A. E. Zuckerman/TSA 245c Jean-Paul Ferrero/Auscape 246c Paul Slaughter/TIB; b David B. Fleetham/OSF 247tr Flip Nicklin/Minden; b MC 248c Kenneth C. Balcomb III/EarthViews 248–249t P. & G. Bowater/TIB 249c David B. Fleetham/IV; b Rick Strange/TPL 250c Kyoichi Mori/OWA 250–251t TIB 251c Kyoichi Mori/OWA 252b MC 252–253t Takashi Yamaguchi/Q Photo International Inc. 253c MC; b MC 254c M. P. Kahl/Auscape; b Steve R. Burnell 255c Jean-Marc La Roque/Auscape; b Flip Nicklin/Minden 256c Clive Bromhall/OSF 256–257t Tony Karacsonyi, Sydney 257c Jean-Paul Ferrero/Auscape 258c Michel Nolan/TSA 258–259t Richard Smyth/Auscape 259c Ken Stepnell/TPL; b François Gohier/Ardea 260c Flip Nicklin/Minden 260–261t Richard Smyth/Auscape 261c MC 262c Steve Dawson/Hedgehog 262–263t Kim Westerskov/OSF 263c Dennis Buurman/Hedgehog 264c Pete Atkinson/PEP 264–265t Pete Atkinson/PEP 265b Peter Hendrie/TIB 266t MC 266–267c Flip Nicklin/Minden; b Harald Sund/TIB 268b Doug Allan/OSF 268–269t Doug Allan/OSF 269c Dominique Braud/TSA 270c Marc Webber/PEP 270–271t Colin Monteath/Hedgehog 271c Ben Osborne/OSF 272–273 Paris, Musee du Louvre/Lauros-Giraudon

Jacket Credits

Front cover tl François Gohier/Ardea; tr Doug Perrine/Auscape; c Brandon D. Cole/Ellis Nature Photography; b Brandon D. Cole/Ellis Nature Photography **Front flap** t Bob Cranston; c Dennis Buurman/Hedgehog; bl Martin Camm (narwhal); br Roger Swainston (Cuvier's beaked whale) **Back cover** tl Martin Camm (fin whale & beluga); c François Gohier/Ardea; tr Gino Hasler (skeleton); l BCA (scrimshaw); l Roger Swainston (narwhal skull); bl Jiri Lochman/Lochman Transparencies; br Paul Ensor/Hedgehog **Back flap** tl Marty Snyderman/PEP; tc WO; tr Roger Swainston (Blainville's beaked whale)

Illustration Credits

Martin Camm 22, 68, 70, 76, 126, 131, 132, 135; **Clive Collins** 164, 165 (Maps); **Marjorie Crosby-Fairall** Chapters 7 and 8 page trim; **Ray Grinaway** 51, 65, 119; **Gino Hasler** 19, 60, 66, 74, 84, 87, 89; **Roger Swainston** 149–152, 157, 158, 160–163, 167, 177, 184, 189, 190; **Kenn Backhaus** All map illustrations; **Genevieve Wallace** Resources Directory

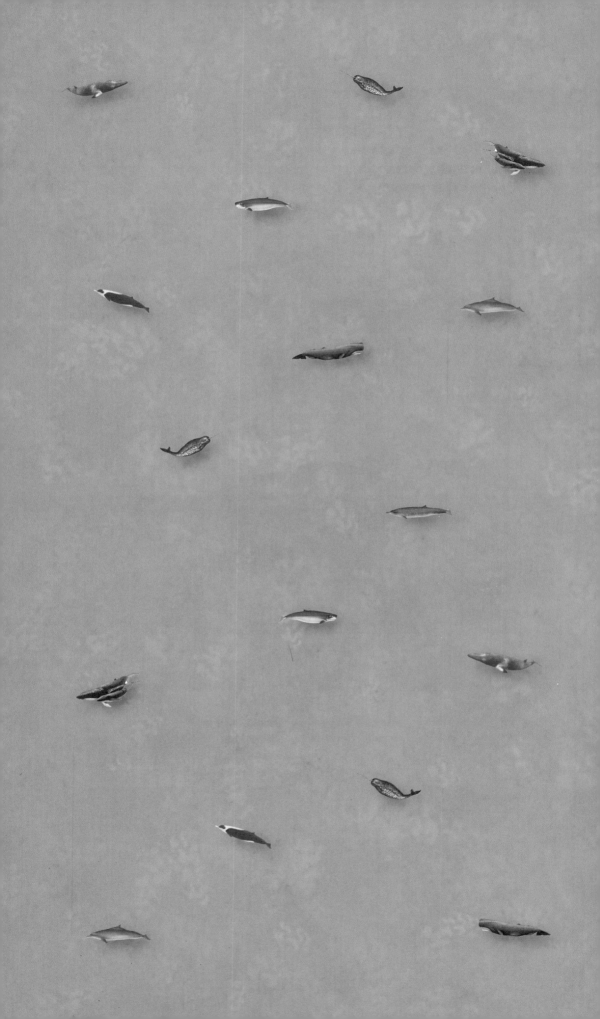